THE NORTHEAST:
A FIRE SURVEY

To the Last Smoke

SERIES BY STEPHEN J. PYNE

STEPHEN J. PYNE

THE NORTHEAST

A Fire Survey

THE UNIVERSITY OF
ARIZONA PRESS
TUCSON

The University of Arizona Press
www.uapress.arizona.edu

© 2019 by The Arizona Board of Regents
All rights reserved. Published 2019

ISBN-13: 978-0-8165-3890-4 (paper)

Cover design by Leigh McDonald
Cover photo by Bob Williams

Library of Congress Cataloging-in-Publication Data are available at the Library of Congress.

Printed in the United States of America
♾ This paper meets the requirements of ANSI/NISO Z39.48-1992 (Permanence of Paper).

To Sonja
old flame, eternal flame

❦

CONTENTS

SERIES PREFACE
To the Last Smoke

WHEN I DETERMINED to write the fire history of America in recent times, I conceived the project in two voices. One was the narrative voice of a play-by-play announcer. *Between Two Fires: A Fire History of Contemporary America* would relate what happened, when, where, and to and by whom. Because of its scope it pivoted around ideas and institutions, and its major characters were fires or fire seasons. It viewed the American fire scene from the perspective of a surveillance satellite.

The other voice was that of a color commentator. I called it *To the Last Smoke*, and it would poke around in the pixels and polygons of particular practices, places, and persons. My original belief was that it would assume the form of an anthology of essays and would match the narrative play-by-play in bulk. But that didn't happen. Instead the essays proliferated and began to self-organize by regions.

I began with the major hearths of American fire, where a fire culture gave a distinctive hue to fire practices. That pointed to Florida, California, and the Northern Rockies, and to that oft-overlooked hearth around the Flint Hills of the Great Plains. I added the Southwest because that was the region I knew best. The interior West beckoned because I thought I knew its central theme and wanted to learn more about its margins. Alaska boasted its own regional subculture. Then there were stray essays that needed to be corralled into a volume, and there were all those relevant regions that needed at least token treatment. Some like the Lake

States and Northeast no longer commanded the national scene as they once had, but their stories were interesting and needed recording, or like the Pacific Northwest or central oak woodlands spoke to the evolution of fire's American century in a new way. I would include as many as possible into a grand suite of short books.

My original title now referred to that suite, not to a single volume, but I kept it because it seemed appropriate and because it resonated with my own relationship to fire. I began my career as a smokechaser on the North Rim of Grand Canyon in 1967. That was the last year the National Park Service hewed to the 10 a.m. policy and we rookies were enjoined to stay with every fire until "the last smoke" was out. By the time the series appears, 50 years will have passed since that inaugural summer. I no longer fight fire; I long ago traded in my pulaski for a pencil. But I have continued to engage it with mind and heart, and this unique survey of regional pyrogeography is my way of staying with it to the end.

Funding for the project came from the U.S. Forest Service, Department of the Interior, and Joint Fire Science Program. I'm grateful to them all for their support. I'm grateful, too, to Kerry Smith, who has edited the entire series with grace and precision and has once again saved me from my worst grammatical self. And of course the University of Arizona Press deserves praise as well as thanks for seeing the resulting texts into print.

PREFACE TO VOLUME 7

WHEN I DECIDED TO survey America's fire regions, I knew I needed—and wanted—to include the Northeast. I just didn't know how and where, exactly, to put it. I fussed with it as a possible minihistory in *Slopovers*. I thought about inserting it as an extended essay in *Here and There*. I wondered why I had thought I needed to give it significant ink in the first place.

The historian in me, however, insisted that the Northeast mattered, and the scholar in me (that training dies hard) demanded as a nod to comprehensiveness that I had to include the region. Because of the unexpected way *To the Last Smoke* evolved, I had not treated the southern Appalachians or the Southeast outside Florida, but I *had* written about Florida, which could somewhat stand for the rest. I had nothing on the Northeast.

My instincts (and conscience) were right. The Northeast holds much of interest. It challenges the Received Standard Version of American fire history and policy. It compelled me to probe events and personalities I had glossed over. It made me reconsider what landscape fire, and its ecology and management, actually means. In the West I had reached a plateau of curiosity: every place began to look like every other place. That was not true for the Northeast. Everywhere I went I learned something new.

My survey was made possible by many people who gently instructed me in how fire worked in their part of the world. I have tried to identify them in the individual essays. But a special acknowledgment must go Lloyd Irland, whose deep knowledge of the region and its forests and whose own survey of fire for the Northeast Forest Fire Protection Compact has established an essential library of statistics and insights. His generosity went well beyond collegial courtesy.

THE NORTHEAST:
A FIRE SURVEY

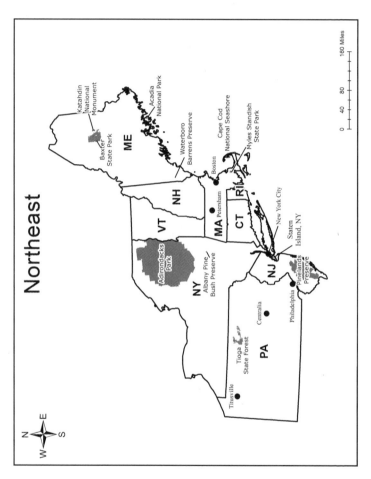

Map of the Northeast

PROLOGUE

Dark Days

O N MAY 19, 1780, the sky over New England darkened ominously. An immense pall of smoke spread from Portland, Maine, to southern New England. It propagated at an estimated 25 miles an hour. "Candles were lighted in many houses; the birds were silent and disappeared; and the fowls retired to roost." For some observers the vision was puzzling; to some, disturbing; to a few, apocalyptic. Was the Day of Judgment at hand? The alarm was such that the Connecticut legislature proposed to adjourn, when Colonel Abraham Davenport rose and declared, "I am against adjournment. The Day of Judgment is either approaching, or it is not. If it is not, there is no cause for an adjournment; if it is, I choose to be found doing my duty. I wish therefore that candles may be brought."[1]

Black Friday was the most celebrated of New England's Dark Days, but it was one of a litany that began in 1716. A notably Dark Day blanketed Quebec on October 9, 1785, and yet another smothered Anticosti in 1814. But smoke was common—northeastern North America had been often discovered by Europeans first by the sight (and smell) of its smokes before shorelines were sighted. Only occasionally did the smoke thicken into vast shrouds that blocked out the sky. They were the pyric foreshadowings of what was to come, first in the Burned-Over District of New England, then farther west. Their rise and fall tracks an era of settlement-inspired big burns. Eventually the Dark Days thinned and migrated west with the frontier, smothering the North Woods around the Lake States and in 1881 sending a pall from Michigan to Massachusetts. By 1910 the Great Fires had moved into the Northern Rockies and merely turned the sun over Boston into a coppery haze. Then they disappeared.[2]

Eyewitness accounts testify to their effects, only rarely, directly, to their origins, but there can be little doubt that the palls of the Dark Days pooled smoke from larger and smaller burns west and southwest of the pall's epicenter, and that some may have spilled out of the Old Northwest. Data from tree rings and fire scars draw a net of burning from the Algonquin Highlands in Ontario to the southern Appalachians to the Missouri Ozarks. A cold front quickened those fires and gathered their smoke, which followed storm tracks between the Great Lakes and the Gulf Stream.

They were, in brief, a premonition, a vast foreshadowing of a wave of fire that rolled westward over the course of two centuries, made fire institutions, and remade fire regimes. The Northeast was a point of ignition for that continental slow burn.

———

Today, the Northeast is a minor feature in America's pyrogeography. Its fire scene has shriveled to a vanishing point. Wildfires, it seems, have gone the way of Pennsylvania's bison. Prescribed fires struggle to become more than boutique burns. A fire infrastructure scrambles to maintain itself, finding closer bonds with Canada than with the rest of the United States. The region could drop out of the national fire scene and likely no one elsewhere would notice.

Yet the story of how American nature and European ideas interact begins here. Even a century ago the Northeast and its mutant progeny, the Lake States, dominated national fire news. America's first great fire of settlement burned across Maine and New Brunswick in 1825; its last, the Maine blowup, came full circle in 1947. Between them the fires that commanded news were those that blasted Wisconsin in 1871, Michigan in 1881 and 1908, and Minnesota in 1894 and 1918. These provided the flame-and-ash vision of ruin to sustain arguments for state-sponsored conservation. National fire policy was as much influenced by the 1903 and 1908 fires in the northern Appalachians and Adirondacks as by the fabled Big Blowup of 1910. The 1896 National Academy of Sciences Committee on Forests that advised on what became the Organic Act for the national forests was staffed with faculty from Yale and Harvard's Arnold Arboretum, as well as consultants like John Muir and Gifford Pinchot. The nation's first forestry schools were located at Yale, Cornell, and Maine. The idea for reserving lands from fire and axe had its model in the Adirondacks Preserve as much as with Yellowstone National Park; the earliest reforms began in states before migrating to the federal government. The Burned-Over District that spawned religions amid the Second Great Awakening later sparked the movement that became a conservation crusade. The 1911 Weeks Act that allowed for national forests to expand by purchase and for federal-state cooperative programs for fire protection was sponsored by John Weeks, congressman from Massachusetts. The first chapter established by the Nature Conservancy (TNC) was in eastern New York, and its first purchased plot, the Mianus River Gorge on the New York-Connecticut border. The concept of land conservation trusts and conservation easements sprang out of the Northeast's peculiar history of land tenure and tradition of local governance. The wilderness movement had its prophet in Henry Thoreau, wandering from Walden Pond to the Maine Woods, and its political realization through a Howard Zahnizer inspired by the Adirondacks. The planning for what became the National Cohesive Wildland Fire Management Strategy occurred at Emmitsburg, on the Maryland-Pennsylvania border. Even American fire literature originated here, with the accounts of Indian burning, the symbolic sermons of Jonathan Edwards, the novels of Fenimore Cooper, and the poems of bonfires and blueberry burning from Robert Frost.

Folk heroes like Ed Pulaski might devise new combination tools, but the educated elite set policy, and they hailed from the Northeast. Like the mountain men who created the Rocky Mountain fur trade, they were not locals, but easterners who learned their trade in the Appalachians and then went west. Jed Smith was from New York, Jim Bridger from Virginia, Hugh Glass from Pennsylvania, and Tom "Broken Hand" Fitzpatrick from County Cavan, Ireland. So, likewise, Teddy Roosevelt came from New York, George Perkins Marsh from Vermont, Charles Sargent from Massachusetts, Gifford Pinchot from Pennsylvania with a career trek to the British imperial forestry school at Nancy, and Bernhard Fernow from Prussia. Conservation did not spring indigenously out of the West, or from westerners going east, but from educated easterners who went west or ventured to Europe. James Fenimore Cooper, of Cooperstown, created the archetypal westerner of American literature, Natty Bumppo—pathfinder, deerslayer, Old Trapper—and set *The Pioneers* in the wilds of backcountry New York. Owen Wister, a Philadelphian, culminated that tradition, now gone Far West, with *The Virginian* (the title alone says a lot). The foundational work of American landscape art, Thomas Cole's *The Oxbow*, featured the Connecticut River painted by an émigré from England. The renowned *Twilight in the Wilderness* by Frederic Church (of Hartford, Connecticut) celebrated the Adirondacks and became the reference for later Western sublime. Of the greatest landscape artists of the Wilderness West, Albert Bierstadt was born in Prussia and Thomas Moran in England.

Those Progressive-minded easterners tried to apply what they knew from learning and experience. In the Northeast the strategy worked. Climate, land use, laws, new technologies, and invented institutions managed to snuff fire out of the scene, as it had in temperate Europe. That model failed when applied to the West, as it did throughout most of Europe's colonies, but the West still bears the scars of the failed experiment. Had those easterners not tried their experiment, westering Americans would likely have slashed and burned through what the country later regarded as sacred landscapes. But the experiment was flawed, and it left a different legacy of landscapes. If it prevented what Teddy Roosevelt called "scalping" the land, it left many of those lands profoundly unsettled. In solving one problem, it created others.

That storied past is relevant for appreciating why America's fire scene today looks like it does. The institutions that oversee American fire were devised in the Northeast; institutions for the federal public domain, institutions for state lands, institutions among the states themselves, even institutions across national borders. Trying to understand today's pyrogeography from the record of the past half century, almost all on public lands, mostly in the Far West, or on agricultural lands, mostly in the Great Plains and the Southeast, is like explaining the evolution of life on Earth without reference to the Cambrian extinction, dinosaurs, or the antiquity of fire. The role of the Northeast is encoded in the institutional DNA of America's fire establishment. That alone makes the Northeast worth reconsideration.

But the region also serves as a counterexample to the West because it shows how an anthropogenic fire regime behaves, because it causes us to question common assumptions about the interaction of climate and people, and because it challenges anecdotal history based on experiences in places where fire is a problem today. Most of America's landscapes are overtly cultural. How should we manage them? To what ends? By what means? And with what kinds of fire? Most landscapes knew fire in their history, today they know mostly fire's absence. What fire, if any, should be restored? The West can remove people and still have fire. The Northeast, largely, cannot.

But that does not mean fire has no place in the ecology of the northeastern biota or that the Northeast has no significance to the national fire narrative, even if, over the past century, the region has mattered more to fire than fire to it.

The Dark Days of smoke palls that blotted out the sky are gone. The region seems to have retreated from the national fire scene into a lair safe from prying eyes, drought, and wind. Where there is no smoke, there is no fire. Yet a better allusion would be to protective coloration. The region still burns prodigiously, but its fires have been invisible because they are burned in special combustion chambers, not free-ranging over landscapes.

Here, too, the Northeast led a wave of burning. Industrial combustion entered the United States through New England; the nation's transition from burning living landscapes to burning lithic ones was field-tested and then fueled nationwide by the fossil biomass of the Northeast. This burning is mostly beyond the field of vision of the region's population, which live in cities or reside in suburbs and exurbs for which power mostly means fuel oil, electricity, gasoline, and diesel, most of whose emissions merge imperceptibly into the atmosphere. Acid rain, pollutants, and greenhouse gases, most of which are invisible to the naked eye, do not darken the sky; smoke's particles cluster around the wavelengths of visible light and are impossible not to see. But they do loom over and foreshadow a kind of darkness.

The region's new dark days reside not in the sky but amid its biota, an ecological darkness caused by the absence of fire. The second nature that human artifice, catalyzed by fire, had made from nature's bounty had morphed into a third nature shaped by the burning of fossil fuels. The Dark Days of colonial America, whose smoke blocked out the sun, had yielded to the dark days of Anthropocene America, whose emissions were trapping sunlight and turning the planet into a crockpot and obscuring the future. What its conflagrations say about frontier America, its missing fires say about postmodern America.

The region's role in understanding the nation's fire past is clear; but is it also pertinent to understanding the troubled future? Most research into landscape fire has focused on the American West, and parts of the East and Southeast where the federal government (the primary sponsor of research) manages the public domain and reserves lands in a wild state or at least a simulacrum of the wild. Most of America's landscapes, however, are in private hands and overtly cultural. They deserve fire management, too. Some still use fire for agriculture and plantation forestry, these in places where a tradition of burning never extinguished. Others probably need some fire restored, or need a fire regime reexamined. Such lands constitute the bulk of the national estate. How should we manage them? To what ends? By what means? What role might fire play in the drama of their existence? Here, once again, the Northeast can prove instructive.

Twice the Northeast has pioneered American fire history, first with axe and torch, and then with dynamo and combustion engines. The first era led to ideas and institutions of national significance for how America would address its landscape fires. Perhaps the second will do something similar for the Big Burn kindled by a fossil-fuel civilization. Whether by its presence or its absence, free-burning fire remains relevant to the ecology of the Northeast. The same can be said for the region in America's national fire narrative.

A SONG OF ICE AND FIRE—AND ICE

THE TEXT BEGINS AS a blank page, a vast sheet of white ice. The ice resulted from the interaction of Earth and Sun, mediated by the distribution of lands and seas and the planetary wobbles and tilts, synthesized into Milankovitch cycles, that determined how sunlight struck them. The ice erased what had been written before, though never completely. It could mold, though not remove, some hard rock; it could widen and deepen, though not flatten or fill, some gorges; it could cover or expose the continental shelf, though not remake its contours. The ice was plural, not singular: it came and went repeatedly across northeastern North America. Each time the ice receded, it left a geologic palimpsest, a hard parchment of terrain roughly organized into inland mountains, coastal plains, and median piedmont of rolling hills, along with scribblings and blotches of outwash sands. Here was first nature, the Northeast as a patch of planetary Earth.

But even as the ice broke, melted, and retreated, and the ocean rose, life appeared along the frontier, moving northward, a slow scramble to claim swathes and niches. People, too, were among that throng. Perhaps as early as 12,000 years ago, they prowled and probed along that border and helped the in-filling behind it. They interacted with the biota, perhaps profoundly in the case of megafauna, until over the millennia, they expanded their interactions to reshape the mobile features of the varied scenes bequeathed by the ice. A world of life replaced the abiotic world of ice. With life came fire, and catalytic fire made possible more interactions and widened the reach of the human hand. Humanity's impact quickened over the past millennia, then accelerated and broadened over the past 400 centuries. People moved soil and sand, dammed and diverted streams, killed wildlife and felled forests, replaced a wild biota with a domesticated one, and they burned. People mediated, and fire enabled. This was a world of second nature.

Then that, too, felt a new fundamental force not known on Earth before. Humanity, the planet's keystone species for fire, in its search for greater firepower, changed its combustion habits. Mostly that burning has occurred in special chambers in one form or another of combustion engines. It began to burn lithic landscapes rather than living ones. The effects have cascaded through the Earth system until, in sum, they threaten to rival that of the ice. They have begun to challenge even the Milankovitch cycles that have powered the ebb and flow of ice throughout the Pleistocene. Most of that geologic era—roughly 80 percent—has been glacial. The last 10,000 years, the Holocene, have been an interglacial, an unusually long one, primed to flip back into ice. The Little Ice Age may in fact have been a tremor heralding the return of continental ice sheets.

It didn't happen. Instead, the Earth began to warm. Human meddling reached gargantuan proportions, attaining in the eyes of many observers the stature of a geologic epoch in its own right, an Anthropocene. The release of greenhouse gases from the burning of fossil fuels has, in particular, stalled the tidal shuffle between glacial and interglacial. An internal combustion era—give it the acronym ICE—may be holding back the old rhythms. But like ice, fire now has assumed positive feedbacks. The Pleistocene ice ages are being replaced by an Anthropocene fire age, a Pyrocene. The upshot might well be called a third nature.[1]

––––––––––––––

Look more closely at that fire scene and its history.

As a fire province, it most resembles northern Europe, also an outcome of continental glaciation. There is a boreal portion—Fenno-Scandinavia, Maine—that has some natural fires. There are sandy zones that favor burning, largely along littorals, but in prominent patches elsewhere—the Landes and the Baltic lowlands, the coastal plains from New Jersey to

Cape Cod to southern Maine. But lightning kindled more barns than snags. Fire-adapted species like pitch pine spread as fire promoted them, which depended on regular ignitions, which had to come from people. Like temperate Europe, the Northeast was an anomaly in a fire planet because there was little climatic basis for fire.

Mark Twain famously commented that if you didn't like New England's weather, wait 15 minutes, and people hold forth endlessly about the changing seasons. From a fire perspective, however, the climate is remarkably uniform because it lacks a rhythm of wetting and drying. Fire must squirm into seasonal cracks; spring after a dry winter and before green up, autumn after dormancy and before snows, especially during sunny Indian summers. Flames interacted with blowdowns, with the once-a-century hurricane, with ice storms, with beetles, blister rust, and gypsy moths. Throughout it all, the forest has proved remarkably durable, but also equally in a state of continual churn. There is plenty of fire in regional history, though constrained, one of many disturbances. Again, as with temperate Europe, fire became prominent when it was put there by humans who slashed, drained, and kindled. The character of human settlement set the character of landscape fire.

What was fire like in precontact times? The most likely regimen would be that typical the world over. Aboriginal economies of hunting, fishing, and foraging display lines of fire—routes of travel burned deliberately and accidentally—and fields of fire, burned to improve hunting habitats, to help cultivate berries, to improve fields of vision. And here, as around the world, one should not overlook the effect of accidental or careless fire, what we might call fire littering.

How such patterns express themselves depends on the capacity of the landscape to carry fire once lit. The sandy coastal plains, rich in pyrophytes like pitch pine and oak, could (not surprisingly) burn easiest. The hummocky piedmont burned patchily by place and season. The absence of a grassy understory, the abundance of wetlands, the broken terrain, all would require repeated, site-specific burning that would leave a fire-mottled landscape. The mountains and coniferous northern forest burned even more selectively, leveraged by occasional lightning fires. In most years, fires would spread haltingly; in those years that opened wider windows for burning, those point ignitions could spread, sometimes explosively. What humans did in such landscapes was not underburn on the southeastern pattern, but leave spot fires that ensured that, when those exceptional conditions for big fires were right, there was always ignition present. There were always campfires and smudge fires unattended along rivers, along trap lines, around busy trails. It's the pattern that characterizes fire in Canada's and Eurasia's boreal forests, and it is hard to imagine why it would not have applied to New England's.

But there was an agricultural fire economy, too. New England was the pointed end of a frontier of maize cultivation that had entered hundreds of (perhaps a thousand) years before European contact. The consequences for fire were both direct and indirect, as with swidden the world over. The direct effects were to expand the realm and to recode the pulses and patches of routine burning. Fire flared where it would not have under natural or aboriginal conditions; by kindling for different purposes, it burned a different regimen on the land. Obviously, as with hunting and foraging sites, some places were better disposed than others. Places susceptible to swidden assumed the fire equivalent of being moth-eaten. Fires returned on the order of decades.[2]

The indirect effects came by establishing where people lived and how they interacted with the surrounding countryside. Near environs would be stripped of dead wood (gone to feed cooking and warming fires). Hunting and gathering would cluster in new ways, which would rearrange the mosaic of landscape burning. River bottoms, for example, would be fire-farmed; uplands, burned in patches to improve travel, gathering, trapping, and hunting. Of course there were limits inherent in the larger setting. Fire could not burn through extensive wetlands, could not propagate where surface fuels like mosses in boreal woods were relentlessly wet, could not race through canopies where the forest was deciduous. The particulars can be hard for later generations to confirm when fallen leaves were burned in the autumn or where surface burns could not generate sufficient charcoal to settle into ponds or where trees scarred by fires were cleared away during a more aggressive wave of settlement. Still, there are accounts from contact times that fit nicely into the above scenario. Those classic accounts have also prompted classic controversies, and since the suite of texts is not adequate to satisfy the demands made on them, they have been subjected to the scholasticism of endless parsing, glossing, and deconstructing, all interpreted through the prism of the viewer's values.

Begin with two reports that describe coastal Massachusetts. In his 1634 *New England's Prospect*, William Wood noted:

> For it being the custome of the Indians to burn the wood in November, when the grasse is withered, and leaves are dryed, it consumes all the underwood, and rubbish, which otherwise would over grow the Country, making it unpassable, and spoil their much affected hunting: so that by this means in those places where the Indians inhabit, there is scarce a brush or bramble, or any cumbersome underwood to bee seene in the more champion ground.[3]

("Champion" is a corruption of "champaign," a derivation from the Latin *campus*, which refers to an open plain.) Such descriptions were the norm along the coast. Thomas Morton elaborated in his 1637 *New English Canaan*:

> The Savages are accustomed to set fire of the Country in all places where they come; and to burne it, twize a year, vixe at the Spring, and the fall of the leafe. The reason that mooves them to doe so, is because it would other wise be a coppice wood, and the people would not be able in any wise to passe through the Country out of a beaten path. . . . The burning of the grasse destroyes the underwoods, and so scorcheth the elder trees, that it shrinkes them, and hinders their growth very much: So that hee that will looke to finde large trees, and good tymber, must not depend upon the help, of a wooden prospect to finde them on the upland ground; but must seek for them, (as I and others have done) in the lower grounds where the grounds are wett when the Country is fired.[4]

(*Savages* here derives from the Latin *silva*, "tree" or "woods," by way of French, and refers to people who live in the woods rather than amid cultivated fields on the European model.)

There was enough burning going on that the newcomers had to burn around themselves "to prevent the Dammage that might happen by neglect thereof, if the fire should come neer those howses in our absence." Morton thought that the "Savages by this Custome of theirs, have spoiled all the rest [of the countryside]: for this Custome hath been continued from the beginninge." Yet he also confessed that "this Custome of firing the Country is the means to make it passable, and by that meanes the trees growe here, and there as in our parkes: and makes the Country very beautifull, and commodious." Writing about New Netherland in 1656, Adrien van der Donck reported that removing those fires had caused rapid reforestation, without which there would be "much more meadow ground."[5]

And farther inland? Peter Kalm, one of Linnaeus's Apostles, noted how indigenes around Lake Champlain were "very careful" about escape fires, yet also observed that elsewhere "one of the chief reasons" for the decrease in conifer forest was "the numerous fires which happen every year in the woods, through the carelessness of the *Indians* [italics in the original], who frequently make great fire when they are hunting, which spread over the fir woods when every thing is dry." Burning, that is, occurred where it could but was not indiscriminate, save where carelessness prevailed.[6]

Or consider the observations of Timothy Dwight IV, president of Yale College, describing expansive "barrens" in western New York, whose "peculiar appearance" he attributed to the fact that the "Indians annually, and sometimes oftener, burned such parts of the North American forests as they found sufficiently dry." Southern New England, "except the mountains and swamps," were covered with oak and pine, well adapted to such burning. The "object of these conflagrations was to produce fresh and sweet pasture for the purpose of alluring the deer to the spots on which they had been kindled." Dwight's QED came when he observed the consequences of removing those annual burns. "Wherever they have been for a considerable length of time free from fires, the young trees are now springing up in great numbers, and will soon change these open grounds into forest if left to the course of nature." He himself had witnessed many such examples.[7]

It would seem that the canonical version of New England history, that the forest returned after clearing by Europeans, may need an earlier epicycle. The forest also returned after the vanquishing of the indigenes and the extinguishing of their fires.

———

Then European settlement created an undeniable wave of flame that, over several centuries, washed across the region and beyond. It helped that the immigrants came from climate, soils, and a biota similar to what which they encountered in America. They brought with them flora, fauna, diseases, fire practices, and an agricultural economy already preadapted to the conditions they encountered. In places—around the colony of New Sweden, for example—they fused with indigenous practices to create frontier hybrids. Colonists learned "fire hunting" from the indigenes. They added landscape draining to slash-and-burn cultivation, and herding to hunting. After the first pass of pioneering, fire flourished within an agricultural matrix that encompassed most of the Northeast.

European contact widened, deepened, and quickened the presence of fire. Fire technology and practices persisted, with some adaptations. Thanks to diseases and wars, the indigenes shrank. More land became available, and more was brought into production. But while intensifying burning and stressing the landscape, the fundamentals of fire ecology remained the same. The big reform was the introduction of livestock, which replaced fire hunting with fire herding, and so encouraged a different regimen of burning. Pioneers pushed fire where, by nature or indigenous economies, it would not have existed, and they pulled fire from where it had routinely flourished. They recoded the pulses and patches of burning to fit a broadly frontier society, and then an agricultural one. Those fires—like settlement overall—passed through regional history like a flaming front.

Again, there are plenty of accounts, and even the origins of a fire poetry. But the most apt might be those of Henry Thoreau, writing at the height of clearing, and just as the region was inflecting into an epoch of depopulation and reforestation, and equally important, as the locomotive was becoming the fire engine of choice.

A hearth fire, he thought, was the "most tolerable third party." But he could see landscape fire with similar empathy. His summary view he wrote on June 21, 1850, two months after accidentally setting 100 acres of Concord woods on fire. Fire, he mused,

> is without doubt an advantage on the whole. It sweeps and ventilates the forest floor, and makes it clear and clean. It is nature's besom. By destroying the punier underwood it gives prominence to the larger and sturdier trees, and makes a wood in which you can go and come. I have often remarked with how much more comfort and pleasure I could walk in woods through

which a fire had run the previous year. It will clean the forest floor like a broom perfectly smooth and clear,—no twigs left to crackle underfoot, the dead and rotten wood removed,—and thus in the course of two or three years new huckleberry fields are created for the town,—for birds, and men.[8]

He pondered the interaction of Indian fires and pitch pine (his favorite). He documented the endless sources of ignition in the cultivated countryside around him—the smoker, the sportsman, the debris burner, the campfires carelessly tended or abandoned, boys collecting sassafras, farmers firing meadows, fallow fields, postharvest stubble, and brush, and increasingly the feral sparks of locomotives. He observed the prime time for fugitive flames to spread was spring, from mid-March to mid-April, when the leaves and grasses were dry, the unleafed trees let sun and wind pass freely, and dry cold fronts blustered through. He recorded how, for severe fires, "the men should run ahead of the fire before the wind, most of them, and stop it at some cross-road, by raking away the leaves and setting back fires." He noted the benign effects of fires, how "surprising how clean it [fire] has swept the ground, only the very lowest and dampest rotten leaves remaining," how "at first you do not observe the full effect of the fire," "how the trees do not bear many marks of fire commonly; they are but little blackened except where the fire has run a few feet up a birch, or paused at a dry stump, or a young evergreen has been killed and reddened by it and is now dropping a shower of red leaves." As for burning meadows, "I love the scent. It is my pipe. I smoke the earth."[9]

More and more, however, as former farmers decamped and the fields fell into the deep fallow of reforestation, the engines of third nature replaced the torches and campfires of the old order. The bad fires were those that followed railways, the Northeast's new lines of fire. In reflecting on the wildfire he had kindled, Thoreau rationalized that he "had done no wrong," and that "now it is as if the lightning had done it," for the fires were consuming their "natural food" as deer fed on browse. Besides, "the locomotive engine has since burned over nearly all the same ground and more, and in some measure blotted out the memory of the previous fire."[10]

And that has been the effect of industrial combustion overall. It has remade the countryside and erased the folk memories of the anthropogenic burning that had once helped power it.

———————

Though condemned by European agronomists and foresters, fire flourished among folk living on the ground. In 1878 Franklin Hough gathered accounts of fire practices from resident observers for his *Report Upon Forestry*. Then C. S. Sargent plotted the pyrogeography for his 1880 census report on forestry, which showed the Northeast holding its own amid the other fire provinces of the country.

Meanwhile, a Great Depopulation was sweeping over the region that left half or more of the landscape fallow. First the Erie Canal in 1825 then the railroads—both pushed and pulled demographic change. Transport made it easier to leave for better lands in the Ohio Valley and around the Great Lakes and cheaper to import farm products than grow them internally. From the mid-19th century to the mid-20th the process steadily drained population away from the countryside, then accelerated in the post–World War II era. In 1880 New England still had over 200,000 farms on 21 million acres. By 1940 those numbers had dropped to 135,000 and 13 million. By 1970 they had plummeted to some 20,000 and 5 million.

That left an ecological vacuum. Trees and shrubs once crowded into margins—white ("old-field") pine in particular—claimed the untended landscape. Fuels held in check by close cultivation and cropping by livestock overran the countryside. By the end of the century those woods had grown sufficiently to be harvested, which prompted another wave of fire, this time wildfire gorging on the slash. Industrial slashing and burning in the mountains left denuded hillsides, which invited erosion and flooding.[11]

By now, inspired by such regional voices as Charles Sargent, George Perkins Marsh, and Gifford Pinchot, a doctrine of conservation argued for change. It was impossible to halt the logging, but it would be possible, given the region's fire environment, to stop the burning. Reflecting the progressive thoughts of the times, Sargent wrote that "fire threatens the forest at every stage of its existence, and a fire may often inflict as much damage upon a fully mature forest ready for the ax as upon one just emerging from the seed; and, as long as such fires are allowed to spread unchecked, there can be no security in forest property." Fire, he thundered, "is the greatest enemy to the American forest"—a conclusion repeated by Pinchot in his *Primer on Forestry*. The threat was no less apparent in his home state, Massachusetts. "Any attempt to improve the forest of the State is useless until they can be secured greater immunity from fire."[12]

Government at all levels made fire protection a priority. Some states like New York created nature reserves to ban fire and axe; some like Massachusetts gazetted state and town forests; some like Vermont and New Hampshire invited the U.S. Forest Service (USFS) to acquire cutover lands in the mountains. Maine established a Forestry District to oversee its unsettled backcountry. The Weeks Act established federal grants to states to help. Its successor, the Clarke-McNary Act, expanded the range of watersheds available to the program. The fires seeped away.

But not before some of the worst wildfires in regional experience broke out. Rail opened up mountain forests and the returned forest across abandoned lands encouraged another round of logging that left slash not seen since 18th-century landclearing; fires followed. The fallow land made for feral fires. Wild fire replaced the tamed fire of agricultural burning. By 1930, however, the era of breakout burns had largely passed. From 1908 to 1930 Massachusetts averaged roughly 40,000 acres burned a decade; by 1990, that fell to under 5,000. Vermont went from approximately 3,000 acres a decade to 300. New York burned 47,000 acres in the 1920s, and 2,500 by 1990. Codes regulated folk burning, agriculture found alternatives to field fires, locomotives ceased to spray sparks, fire protection systems toughened, land use spun away from fire-catalyzed occupations—all the usual prescriptions that work in temperate environments in which natural fires are rare came into play with reductions on the order of 10–20 in burned area.[13]

Increasingly, the Northeast moved from visible flames to remembered ash. Big burns (by regional standards) occurred after 50- or 100-year events like the 1938 hurricane (in which windfall replaced logging slash) or the early 1960s drought. The last great aftershock came, unexpectedly, in October 1947 along the coastal plain, mostly in Maine. That outbreak inspired a regional consortium among the states, the Northeastern States

Forest Fire Protection Compact. What had been a national hotspot in 1880 became by 1980 a cold ember, quickly overgrown by new woods, its pyric history all but forgotten.[14]

Most observers felt about the banished fires as today's residents would the obliteration of Lyme disease. They were happy to remove fire from landscapes and stuff it into machines. Only later would the ecological effects of fire's wholesale removal slowly become apparent. Pine and oak suffered; wildlife suffered; shrubs like blueberry and rare forbs like poor Robin's plantain suffered. The fires that had animated those scenes did not have to come from nature to be significant: they just had to come. They were the fires of second nature, not first nature. When they no longer came, the lands that had long known them sickened. The flames they had depended on were now shackled in combustion chambers. A phase change had swept over the region's pyric history.

Its celebrated fire history had remained within the realm of second fire. Much as the natural landscape had been remade into a humanized second nature, so had fire. Second fire (as it were) encompassed all those fire practices humans distributed about the land, particularly those for which fire served as catalyst; people did not burn just to burn but as part of how they lived on the land. By the onset of the 20th century the fire regimes of the Northeast bore little resemblance to what might have existed under purely natural conditions. By the end of the 20th century, those fire regimes—for that matter, landscape fires—were mostly gone.

———

This time the shift did not represent changes and mutations within the realm of second fire, but the wholesale replacement of second fire by another realm of combustion, the burning of fossil fuels. More and more, the Northeast burned lithic landscapes instead of living ones. The region had inflected into an age populated by machines that fed on fossil fallow, an era aptly symbolized by the internal combustion engine. A new ICE age spread over mountain, piedmont, and coast. Feeding on fossil biomass it freed up forest biomass. The ICE age of the Anthropocene promised to refashion the region as the Pleistocene's ice age had before.

Its anthropogenic fires had favored some biotic elements over others: it shuffled species and recycled biotas, but it had not destroyed them.

People used a natural process to reshape natural materials. Second nature was first nature refashioned through the artifice of humanity, but through means and tools themselves remade from natural sources. Third nature reworked second nature, and it did so with means unlike anything in first nature, or with processes so reduced and isolated that they no longer did ecological work, even as surrogates. Third nature is a built landscape typically fashioned or framed with asphalt, concrete, glass, steel, plastic. A stone fence was made from rocks removed from a field and relocated. A skyscraper was made from cement and metal burned from stone and ore that bore no interaction with its quasi-natural setting. The same has held for fire. Fire codes increasingly ban any form of open fire, even leaf burning. Open burning went from being plowed under to being paved over.

In 2009 David Kittredge wrote a commentary for the *Journal of Forestry* that spoke to the "fire in the East." He meant it metaphorically. The megafires of the West that drew public attention "do not *destroy* forests," but the sprawl that characterized contemporary land use in the East did. "Forests do not grow back after development." Through millennia of human use, through four centuries of European-style land conversion, the Northeast's forests had endured. Second nature had preserved forests, or where cleared they had retained the capacity to return. Third nature was different. Fire history was reifying Kittredge's allusion. Third nature had its own combustion, not drawn from or dependent on the character of its surroundings. It was more ruthless, and potentially more ruinous, because it could deny the ability to rebound that had characterized the Northeastern forests since the ice had left.[15]

———————

The fire story of the Northeast is the fire story of temperate Europe transplanted to the New World. But just as northern Europe had been the hearth from which European peoples and norms had birthed a second age of colonization, and from which modern science and the industrial revolution had been disseminated around the world, so was the Northeast for North America. What happened in the Northeast affected national fire norms and institutions.

That transfer of understanding led to many errors. Temperate Europe is not normative in fire-planet Earth; the Northeast is exceptional in fire-prone North America. The assumptions that what made sense in a region that lacked a climatic basis for routine fire would make sense in places like the Far West that had abundant natural fire, or like the Southeast that had deep traditions of cultural burning, were flawed. The 20th century showed how erroneous this presumption could be.

But there is a flip side to this story. However inappropriate its understanding of fire might be outside the Northeast, it suits the region itself. The fire practices of the Northern Rockies, or the Wichita Mountains, or the red hills of Florida can be equally maladapted in the Adirondacks or White Mountains. There is clearly a place for fire, but it will not be justified by appeal to wild nature or prescribed fire as an informing principle of land management. Fire will be overwhelmingly anthropogenic; it will occur in second nature, or be restored amid third nature; it will flare or subside within a philosophically awkward pluralism of peoples, purposes, and practices. The Northeast has its own fire rationale. It deserves a history that builds on its reality.

=========

It's a place of mixes. It has long mixed people and nature. It has a mixed economy, a mixed forest, a mixed history, mixed land tenures, mixed aesthetics, mixed governance. Its history is kaleidoscopic; the pieces seem to endure, or are metamorphosed in the way exurbanites reclaim colonial farmhouses or a software company repurposes a historic building; then they recombine. There are private lands with public purposes; there are public lands owned by private interests. Donations, land trusts, and conservation easements do what the federal lands do in the West. The grand public domain that has animated so much of American environmental philosophy and politics is absent, or one small voice in a choir. Instead of preserving wilderness, the Northeast is watching significant patches of its estate rewild.

To an outsider it can seem bewildering, a tangled bank of historical legacies and modern ambitions. What works in Montana or Florida doesn't in Massachusetts or New York. It's a managerial mashup. That holds, too, for the region's fires, which after all must synthesize their surroundings. In parts of the public-land West, history can seem irrelevant. It's enough to manage fire through first principles in the here and now.

In the Southeast history matters—the past is not even past, as William Faulkner put it—but through continuities in practice, particularly the tradition of anthropogenic fire, as with the oft-repeated stories about learning fire by helping a grandfather burn land. In the Northeast the scene can only be understood through a historical palimpsest of landscapes and ideas and institutions. The region has been remade by ice, then people, then newly (technologically) empowered people. The landscape story is one of change, resilience, and pluralism. So, too, is the way that history itself might be understood.

Today, outside southern New Jersey, the region has few significant fires and no fire crises that seem destined to influence national policy. The concern among the fire community is deciding what fire, if any, to reinstate amid the continual churn of land use. But just as the region's fire scene cannot be understood without understanding its history, so the history of the American fire scene nationally cannot be appreciated without the Northeast. You don't need big fires to have a fire history. You don't need conflagrations to have a fire problem. You only need to have people and nature interact in ways fire mediates. In the Far West fire history is in your face. In the Northeast it's more likely in your pocket.

WHERE THE PAST IS THE
KEY TO THE PRESENT

T HE HARVARD FOREST IS as famous a landmark in forest science
as Walden Pond is in literature. Operated by a university without
an undergraduate forestry school, embedded in an undisguisedly
cultural landscape, not a part of the public domain that has long been the
core of American forest thinking, dedicated rather to working landscapes,
it may be the premier forest research site in the country. It is as important
for understanding fire in the Northeast as Tall Timbers Research Station
is for fire in the Southeast or Konza Prairie in the Great Plains. But Tall
Timbers was chartered to study fire ecology, and fire is indispensable for
tallgrass prairie; the Harvard Forest included fire because fire was, for a
time, around and, shortly after its founding, significant. Its true informing theme is history. Aldo Leopold famously noted that the first rule of
intelligent tinkering was to save all the pieces. Here the cogs and springs
are historical. Harvard Forest is a long-term ecological research site for
which "long-term" dates back to presettlement times. It's a field station
for landscape history.[1]

The history of Harvard Forest as an institution now spans over a century. Its story begins in 1907 when Harvard acquired 3,000 acres near
Petersham, Massachusetts, as a field lab, research center, and demonstration forest. The refelling of the New England forest was at full throttle
(the cut would reach its apogee in 1910); conservation was becoming a
national crusade; and Cornell and Yale both had forestry schools in much
the way universities today are creating schools of sustainability. In 1914 the

forest was made a graduate school. Later, it added a 25-acre tract of old-growth white pine and hemlock amid what became the Mount Pisgah State Forest in New Hampshire, 100 acres of planted and upland woods at Matthews Plantation at Hamilton, Massachusetts, and the 90-acre Tall Timbers Forest at Royalton, Massachusetts.

Under its founding director, Richard T. Fisher, a Harvard graduate, a former member of the Biological Survey, and among the founders of the U.S. Forest Service, it sought to place forestry on a broad historical and cultural foundation, and so show how forests could be both used and conserved. As with the U.S. Forest Service, and forestry generally, Harvard Forest assumed that forestry would pay for itself, and again like the USFS, its evolution has mirrored a general shift from conservation to sustainability. In 1988 Harvard Forest joined the long-term ecological research program sponsored by National Science Foundation, and later the National Institute of Global Environmental Change program and National Ecological Observation Network.

It was an odd forest reservation because it was an unabashedly cultural landscape, not a place set aside to preserve an intact nature from fire and axe—that made it distinctive from the onset. It told a story of human artifice played out on soils, streams, and woods. So it was entirely appropriate that the history uncovered by the Harvard Forest itself should be expressed through another form of artifice in the form of a series of dioramas. The dioramas made Harvard Forest famous. Whatever controversies attach themselves to the themes investigated, the dioramas provide graphic buttressing. For most of the public the dioramas are better known than the forest.

In the late 1920s Fisher and Albert C. Cline, who later became the museum director, worked out the basic designs. With funding from the philanthropist Dr. Ernest G. Stillman work began at the studios of Guernsey and Pitman, in Harvard Square, in 1931 and ended in 1941. (Fisher died tragically in 1934 at age 57.) Upon its completion the Fisher Museum was dedicated to house them for public education.

The historical suite consists of seven dioramas beginning in 1700, then documenting the changes in the same place for 1740, 1830, 1850, 1910, 1915,

and 1930. Collectively, they trace the whole cycle of European colonization, land clearing, and land recovery. Another suite looks at old-growth
forests, wildlife, soil erosion, and forest. (These displays were developed
with advice from other specialists on the Harvard faculty.) A third suite
depicts various methods of forestry: thinning, pruning, planting, harvesting. Altogether they are a kind of taxidermy for landscape.

Fantastically detailed, eerily realistic, the dioramas distill into a series
of snapshots over 300 years of interaction between changing people and
a changed land. What could take thousands of words of texts and scores
of data in tables and graphs is instantly visible to even the uninitiated eye.
Most people view the dioramas through photographs taken of them. But
however dense the pixels, they don't convey the dioramas' depth. It's their
three-dimensional character that allows you to see the details (the paint
can, the woodpecker) that grant the scene its verisimilitude. Collectively,
they add a fourth dimension. They are cross-sections of history.

The dioramas make the Fisher Museum into a planetarium for land
use history. Almost anyone who studies the region's environmental history will refer to them. They are as famous in their way as such classic
tales of environmentalism and science as the fight over Hetch Hetchy
and Darwin's finches. Here was an example of another kind of second
nature—a second-order nature—in which art has remade our understanding of landscape.

──────────

The first of the series, the Presettlement Forest, lacks a human presence.
The others have, in the diorama's foreground, a person doing what most
shaped the scene. The 1740 pioneer clearing and planting amid dense
woods. The 1830 farmer harvesting in a mostly cleared landscape. An 1850
hunter prowling the prime habitats made by mixed fallow, brush, and pine
reproduction. Two 1910 loggers cutting the now-mature old-field pine.
Then, an anomaly: two deer replace people, as the 1915 landscape is again
abandoned and human population reaches a nadir, and hardwoods seed
instead of pine. Finally comes a 1930 painter, representing a class of "summer people," attracted to the aesthetics of the land. There the dioramas end.

Since then, Harvard Forest has added 87 years, or roughly another
38 percent of the history chronicled. What additional entries might be

included if the dioramas were extended? Surely one would deal with the 1938 hurricane which flattened over 70 percent of the forest's trees. (The Big Blowdown was for New England what the Big Blowup was for the Northern Rockies.) Another would be the contemporary scene in which exurbanites reclaim the old farmsteads into subdivisions, this time not clearing but allowing the woods to thicken and perhaps posting against use of the land by nonresidents. The diorama would need a paved road. The foreground would have people in a car, probably an SUV.

But the greatest omission is the one missing in plain sight. The forest was established to do research. Its prime use is natural science, alloyed with historical scholarship. What is missing is the forest as a research site, with researchers huddled over instruments, as timberjacks earlier worked a felled bole. More remarkably, of the six dioramas that span settlement, the scientist could—should—be present in three. Half the story of Harvard Forest itself is its use for field research and experimentation. What is missing, that is, are the people and the reason the forest exists in the form it has enjoyed since its founding a century ago.

It's a curious omission, suggesting that science transcends the phenomena it examines, that experimentation is not a form of land use, that the observer is not part of the observed nor affects it through observation. (Interestingly, the Heisenberg principle of indeterminacy was announced in 1927, just before the last of the dioramas.) Yet it is clear that the founding managers of Harvard Forest intended to manipulate the site in order to test various methods of silviculture. The Progressive Era was an age of activism, an era when pragmatism was given formal structure, a time of reform. As a demonstration site, Harvard Forest would be shaped to that end.

Whether through modesty, or philosophical naiveté, or a bowdlerized positivism, the most prominent actor and commentator is not included in the scene. This omission has affected not only how the forest is presented to the public but how its inhabitants understand the dynamics and lessons of their land.

━━━━━

In 1966, nearly 60 years after its founding, and 25 years after the dioramas were completed, its then director, Hugh Raup, published a celebrated essay on what the experiment that was Harvard Forest had taught. "The

View from John Sanderson's Farm: A Perspective for the Use of the Land" became the most oft-cited article ever to appear in the *Journal of Forest History*. Here was yet another artifice, this time textual, applied to the land.

The title reflects the fact that the core of Harvard Forest was, during its settlement period, a farm owned by John Sanderson, a successful farmer and businessman. The view from his farm looks back to the past and ahead to a future. When Sanderson was prospering, the view to the future looked like the present prolonged. That didn't happen, for reasons well outside Sanderson's control and events beyond his, or anyone else's, vision. The same scenario happened with Harvard Forest. It was founded to establish a permanent demonstration of conservation-informed forestry, one that would pay for itself. The future went in another direction. What John Sanderson was to the early 19th century, Ralph Fisher was to the early 20th. Sanderson managed to sell out before the big change came. Fisher died early before his vision of the future faltered.

What this meant, to Raup, director of the forest when he wrote his commentary, was that the "principal role of the land and the forests has been that of stage and scenery." The critical factor was "the human element," as synthesized by markets. Raup questioned the received wisdom of conservationists that reserving land for a long future was the essence of good management. After all Sanderson's farm produced "better economic results" than Fisher's Harvard Forest. Raup concluded that "a fundamental problem in modern resource management is finding a way to bring its planning horizons within sight of the people who have capital to invest." The moral arguments for conservation were pointless. Long-term planning by experts (or government) was meaningless. Both strategies were premised on a stability which did not and could not exist. (Certainly the 1825 Erie Canal and the 1938 hurricane showed the unpredictability of future events.) Published two years after the Wilderness Act of 1964 and at the onset of another decade of state-sponsored environmental reforms, Raup's "anti-planning parable" as a critic termed it found wide circulation. In a sense, Raup was the Milton Friedman of forestry.[2]

It did not spark a formal rebuttal for another 50 years, when Brian Donahue published "Another Look from Sanderson's Farm: A Perspective on New England Environmental History and Conservation." Donahue contextualized Raup's "challenge" as "after the heyday of the gospel of efficiency in resource management, but before the rise of the modern environmental movement," and summarized Raup's conclusion as championing "laissez-faire: Get it while you can." He noted, too, that during Raup's tenure as director (1946–67), emphasis had shifted from classic forestry to ecology and history.[3]

Donahue's own challenge began by repudiating the assertion that the land was just there, as so much inert plastic to be molded by bustling humans. Rather, it had its own character and logic that influenced what people might do; and hence it shaped history. He noted a fascinating counterexample to Raup's reliance on ingenuity and markets. Farms, abandoned, had not promptly ceded to old-field pine, but had first converted from arable to pasture, which changed demographics; meanwhile, natural reseeding went apace. The encroaching woods shrank pasture. "Pastures were not filling up with pines because farms had been abandoned. Even on prosperous farms, pastures were abandoned because they were filling up with pines." The stage was, in fact, an actor. The play was the dialogue between people and land.[4]

Moreover, the putative failure of conservation was actually the failure of forestry as a governing doctrine. Much as the land had changed, so had conservation. As fields became pastures, and pastures evolved into forests, so conservation had morphed into environmentalism, and environmentalism into sustainability. The new order reemphasized the land because the land held the cogs and springs, the pieces and processes, that allowed for people to work with it to fashion a future. Stripping the land as modern markets were doing destroyed its capacity to tolerate new options. The future was not knowable, and was certainly not stable, but without some variety of conservation it would be narrower, more impoverished, and less plastic. Stripping the land made it into Raup's empty stage, which ended the play. A Leopoldian land ethic was not a meddlesome irrelevance to the future but an ethos that allowed land to provide ecological goods and services and people to continue to interact with them. It underscored the Wildlands and Woodlands initiative behind the protection of the Massachusetts forest. It tracked the evolution of the Harvard Forest from being a demonstration forest to a long-term ecological research site.

Yet another entry appeared in 2014, this one from David Foster, Harvard Forest director since 1990. He rehearsed the contrasting views of

Raup and Donahue, and then introduced another factor into the layered narratives. For Raup the history of John Sanderson's farm pivoted on the Erie Canal, which upended market relationships: an idea from outside the region altered relations with the region. For Donahue the pivotal event was the reclamation of pasture by pine, which forced farmers to adapt: a process intrinsic to the biota redefined market possibilities. For Foster, it was the hemlock forest that had predominated the woods for 8,000 years and that supplied the raw material for a tannery on the farm. "It was hemlock that made the whole tannery operation work and ultimately allowed John Sanderson, his family, and farm to thrive."[5]

In Foster's reading the tannery was the vital catalyst, the economic asset that moved the farm from sustainable to prosperous. (Sanderson learned tanning early.) It was the income from the value-added tannery that made the farm into a successful business. Unraveling the story of the tannery was a complex enterprise that relied on palynology, ecology, archaeology, and written records, decoded by paleography. If, at the forest, nature and culture fused, so did natural science and humanistic scholarship. Hemlock was a keystone species in the farm's economy that symbolized perfectly the interaction that made the farm what it was. In some ways, it still does, as it confronts extinction in the face of climate change and invasive insects like the wooly adelgid.

This third view is deeper than the others—its pollen dates back 10,000 years. It is also more particularized. It targets one tiny portion of the farm, the switch that turned the dynamo on. What you see, that is, depends not only on perspective but on scale. As microbes matter to regional ecology, so specialized woodlots and grindstones by seasonal creeks matter to human habitation. Without the hemlock, there would have been no tannery; without the tannery the wealth of John Sanderson would not have allowed him to enlarge his holdings; without that expanded farm there would have been no Harvard Forest. At each stage scale was critical. In the farm's economy the hemlock woodlot mattered as much as the Erie Canal; in nature's economy, the old-field (or rather, pasture-infesting) pines mattered as much as loggers.

What has changed is not only the land, but how we understand the changes we see. Each man viewed the scene through the prism of his society. John Sanderson died at age 62 while unhitching an unruly oxen in a barn. Richard Fisher died unexpectedly at age 58 while golfing. Hugh

Raup, having retired at age 67, died at 94 in a nursing home. Each saw the scene through the cultural lenses of his day. Each proposed a parable, seemingly appropriate to his times.

There is little reason to believe that David Foster's third view will be the last. But it is hard to imagine another not occurring on Harvard Forest and not framed by those dazzling dioramas.

———————

None of the classic seven dioramas show fire. It's easy enough to imagine how fire could fit into each, and among the forestry suite there are two fire displays, one showing a firefight underway (overseen by a metal lookout tower) and another, the burned area afterward. Undoubtedly fire was used in landclearing, the burning of straw and fallow, the refreshing of pastures; there were horrific fires amid the slash of industrial logging; there were fires among the shrubs and revanchist conifers. There are fires today amid New England's sands, scrub oaks, and pitch pines. Some of those flames are deliberately set to enhance ecological integrity. But that is not the challenge to the American fire community that the forest poses.

Behind the fire revolution were two fire axioms. One was the vision of fire as natural. People set many bad fires (and many good ones)—of course. But the argument that fire didn't belong because it was a human artifact was effectively hanged, drawn, and quartered. Fire belonged because it was an intrinsic part of the natural world; and through it the goodness of the wild could be brought, by means of prescribed burning, to degraded landscapes. Yet natural fire has almost no place in central Massachusetts. The fires that matter are those set by people. The other axiom was that, where fire had been unwisely excluded, it had to be reinstated to the proper restore point. In the West this has almost always aligned with notions of wilderness. But there is no untrammeled wilderness or obvious restore point at Harvard Forest, no climax state, no angle of biotic repose. The proper role of fire must be set by people. The clarity promised by appealing to nature becomes a cultural compromise, to be negotiated among humans. So, while the dioramas are a marvelous distillation of the forest's history, they don't say what its managers should do now. Even the forest's managers avoid that issue by concentrating on research, which appears to stand apart from normal land use.

So it can seem to casual viewers that Harvard Forest holds lessons for the Northeast, but not for the rest of the country, particularly the vast public-domain wildlands of the West, or that it can inform only if we shift our lens from wildlands to cultural landscapes, and for cultural landscapes natural science is by itself an inadequate method of inquiry. In fact, those western wildlands are also cultural creations, though with different dynamics. They, too, require that other scholarships interact with science to sculpt an understanding, just as people interact with nature to shape landscape.

Pragmatism as a formal philosophy had its hearth in Cambridge, Massachusetts, with the Metaphysical Club and Harvard College philosophers like William James before migrating with John Dewey to the University of Chicago and then Vermont. It's a philosophy of life as experimentation, for a world without fixed pasts and foreordained futures, a perspective in which no one person or group holds the torch of truth, in which pluralism is a given.

Harvard Forest writes this philosophy into land history. It proposes a view of many pasts and many pathways toward nature protection. It suggests that history is not truly cyclic like the rotations of the Earth but idiographic, full of rare disturbances that compound in unpredictable ways. Such a vision subsumes fire science within fire history; it suggests that, with regard to landscape, even a putatively transcendent knowledge is actually grounded in place and time. It's likely that the torch will, in the future, as here and there in the past, light the way of that understanding. And that vision is why Harvard Forest deserves to join American fire's long pilgrimage trail, along with Redwood Mountain, Tall Timbers Research Station, and Mann Gulch.

FIRE'S KEYSTONE STATE

PENNSYLVANIA IS NOT A keystone state in the national narrative of fire, though it once was. Still, its fire scene, even today, can take the shape of its signature logo. It has an arch—the long arc of fire's narrative—and four anchor points. Two deal with fatal fires, one with the foundational story of Pennsylvania forestry, and one with the future of fire in the commonwealth. The assembled outcome speaks to a keystone place, and even more, to a keystone species.[1]

It's a longer, richer saga than generally believed. Most of Pennsylvania lies within the Appalachian Mountains and its backside, the Allegheny Plateau. It's a corrugated, textured landscape full of niches and nooks, not easily contained within a single biome or regimen of burning. Like its central Ridge and Valley province, fire's history has its crests and troughs. But the state's settlement history over the past 150 years has so overturned what preceded it that the contemporary scene has been unwisely projected back into the deep past. Over the previous two decades researchers have broken the myths that fire has no historic or ecological role in oak forests, that Pennsylvania had no fire history before the advent of the axe, that only thick forests, not a medley of barrens, woodlands, and savannas, defined its pre-Columbian biota, that fire was, by humans, unwisely introduced and then, by other humans, out of ecological necessity, removed. Rather, fire extends from the coal seams in Pennsylvania's deep past to the bio-rich barrens of the Poconos and many places and times in between.

FRONTIERS OF FIRE, THEN, AND AGAIN, AND NOW

Pennsylvania has two grand fire provinces. One thrived in the Paleozoic and remains fossilized in stone. The other has flourished after the Pleistocene and burns the living landscapes that filled up the state in the wake of receding ice. Its more recent American fire narrative has two inflection points. The first occurred when Europeans landed and began to remake the land. The second occurred when the fossil biomass from the Paleozoic was reburned and collided with the living biomass of the post-Pleistocene.[2]

=====

The conventional narrative, at least among foresters, has been that fire was not integral to Penn's woods, that there might have been patches of landscape that were regularly burned, but the northern pine and northern hardwoods were far from being fire obligates and the fires that state forestry heroically removed were those that people had wantonly introduced along with weeds and cholera. Yet it's possible to pick up the other end of this storied stick. It may be that a fire-receptive biota was the norm, burning according to varied rhythms, and that fire-intolerant species existed in refugia, only liberated broadly in more recent decades by fire's exclusion.

A pre-Columbian fire history is murky; the sources are sparse, the story complex, the interpretation of data encumbered by disciplinary lacunae and biases. In a humid climate few stumps or snags survive to record a chronicle of burns, and in an oral culture, no folklore or songlines survive if the people perish. Most fire science was developed for the American West, amid an assumption of wilderness, not for overtly long-settled landscapes like those of the Northeast. But fire there was nearly all human caused; nearly all in the spring and fall, outside the growing season; nearly all in landscapes not dominated by shade forests or wetlands.

Patches of pitch pine, barrens rife with berries, oak-shaped woodlands—all served as corridors to carry flame from the fire-flushed coastal plains inland. Most such firescapes survive today in protected refugia, but paleoecology and historic reconstructions suggest they had previously claimed broad swathes, flourished along traditional routes of travel, and sustained a fire province frequently burned by surface fires. The Pocono Plateau once held an estimated million acres of "barrens," routinely fired in spring freshets of flame. The Ridge and Valley region knew fires along its valleys, south-facing slopes, and ridgetops. Pine Creek Valley along the eastern Allegheny Plateau, with oak and red and white pine intertwined, experienced surface fires on the order of decades.

There is little reason to believe that Native American fire practices were different from those throughout the region. The leap is easiest where barrens and pitch pine-scrub oak wended inland, or where tendrils of prairie or savanna lured wildlife through the woods. Foragers burned to prune berry patches and to ease the gathering of chestnuts and acorns, and where rattlesnakes were present to expose their presence. Likely hunters burned in the spring to freshen browse, and in the fall to strip away fallen leaves and encumbering logs to ease a fall hunt. The game was lush, famously so. Even bison seasonally moved across western Pennsylvania. Not until 1760 did the wedge of settlement separate the herd north and south, and not until December 1799 did Pennsylvania bison go extinct.[3]

Still, proxies and analogies narrate a plausible fire history. Charcoal and fire-scarred trees tell a surer one, adding quantitative rebars to the narrative concrete. The oldest reach back to the 16th century.

From the Ridge and Valley region: Fire return intervals varied from one to five years in pitch pine, overwhelmingly in the dormant season, but with a long rhythm of fire-flushed and fire-free eras that seem to correspond to human history. (Lightning fires account for 1 percent of fires, as documented by records of the past 50 years.) The dappled pattern of fire history helps explain the periods of fire lacunae important for the establishment of less fire-resistant cohorts. The sparse growing season fires may be significant for certain species. Terrain and climate are mediating factors. The fire frequency is higher than for red pine farther north, but comparable to white oak a bit south. It's a pattern typical of the southern Appalachians.[4]

From the north-central region: Prior to Europeans the red pine of the Pine Creek Valley (along the eastern flank of the Allegheny Plateau) did not experience permanent human settlement, but knew transients who passed through as hunters and warriors. It was a place to hunt and forage, not farm, and served as something of a demilitarized zone between the Iroquois and Susquehannock. The prevailing fire regime saw "sporadic, large, dormant-season fires separated by decades of no fire." Fires came every 35 to 50 years, only weakly correlated with drought. Coinciding with the advent of European diseases and the Beaver Wars (1650–1735), and again during the years of the American Revolution, fires vanished. In a pattern typical of the northern Appalachians they reappeared, with increasing vigor, as settlement renewed. They subsequently exploded as industrial logging trashed the landscape, then again disappeared with industrial fire suppression.[5]

From the Pocono Plateau: The woods were mostly scrub oak and pitch pine, and the ground cover heath, including swathes of huckleberry and blueberry, the southernmost outlier of a corridor of barrens that skipped north to New Brunswick. Anthropogenic ignition seems to have quickened around 2,000 years ago, allowing the Pocono barrens to expand. Seasonal hunters and berry pickers kept the patches simmering with fire. The Pocono complex may have clothed over a million acres.[6]

The other check on how effective anthropogenic fire may have been is to observe what happens when it is removed. A Pocono Fire Protective Association was organized in 1902. Gradually, hardwoods closed in, the heathlands contracted, biodiversity shriveled. The core Pocono barrens survived because fires continued. (In 1933 Yale forestry professor Ralph Hawley marveled how "fires of incendiary origin intentionally set for the definite purpose of burning over forest tracts apparently are still of too common occurrence.") Thanks to berry cultivation the fires continued into the 1960s. But aggressive fire suppression between the mid-1960s and the mid-1990s led to forest conversion in 73 percent of the barrens. Only routine fire could have sustained such a biome, and only human ignition could have accounted for the magnitude and timing of those fires. Today, a scant thousands of acres persist in anything like their original condition.[7]

Not every site could burn every year, and fire, as always, had to interact with everything else in the setting. Climate worked to ramp up or dampen the power of fire to propagate; drought years widened the range available for fire spread. Wetlands were abundant. Likewise fauna boosted or depressed the fuels available for combustion. Beaver created ponds, bison cropped grasses, passenger pigeons feasted on acorns and by their sheer numbers broke the branches of oaks. The evidence grows that fire was an all-spectrum catalyst especially in oak forests, without which oak struggled to regenerate.

Those interactions were less mechanical, like a throttle on an engine, than a web, where each interaction spilled and rippled into others; and besides, all the parts had their own dynamic and history. There was a time when bison had not reached Pennsylvania. There may have been a period when passenger pigeons were far fewer in number, only erupting into epidemic proportions as the old order, fashioned by American Indians, crumbled and let new oak colonize into the vacuum. Then, that prior world was slashed and burned into near extinction, and a wave of flame consumed the record of what preceded it.

The cultures that burned knew what fire did from empirical evidence. They burned to stimulate blueberries and huckleberries, to promote good hunting grounds, to ease travel. If they fumbled their fire practices, they suffered. Those fires rose and fell as peoples and new technologies came and went. Some of that knowledge transferred to the newcomers where it fused with other pioneering lore. Almost all that knowledge, however, was wiped out during the Gilded Age's Great Barbecue that saw industrial combustion interbreed with open flames. Modern science is still far from reconstructing that lost lore or understanding the landscapes it made possible.

───────

The old order began unraveling with European contact. Europeans established two colonies along the Delaware River, the beachhead for what evolved into Pennsylvania. One was famous for close settlement and careful planning, the other for the free-booting frontier that sprawled west.

The best known is William Penn's Quaker city, Philadelphia. Penn purchased his royal charter in 1681, followed by land purchases from the Lenape tribe and a charter of liberties for prospective colonists. Within a century, bridging agricultural near-hinterlands with maritime commerce, the city became the premier urban center in North America. The lesser known but more powerful colony was New Sweden, established in 1638 (and extinguished by Dutch conquest in 1655). It was a determinedly secular trading post, not a place of sanctuary for religious refugees. It quickly evolved in unexpected directions by intermixing two oddly similar cultures—that of the local Lenapes with émigré Finns who had restlessly pushed around the Baltic. That fusion became the pith of the famous

American backwoods style of pioneering, one further stiffened by Scots-Irish. It was a society as footloose as Penn's planned estate was rooted.[8]

By the time of the American Revolution, there were nesters in the valleys, and rovers on the ridges. The mountain land was widely but lightly settled, the flotsam of a sprawl that kept pushing west, that reached from pitch pine to tallgrass prairie, from the Alleghenies to the Ozarks. A snapshot of the divergent outcomes can be seen from two sites in southeastern Pennsylvania, near their prospective hearths. The Amish of Lancaster County are a dictionary definition of a rooted community. Practically within chimney-smoke distance to the east lies the Daniel Boone homestead, where the frontiersman grew up before following the great valley south and across the Appalachians.

Both nesters and rovers burned. Slash-and-burn settlement was the norm, whether it involved consuming the felled woods in one gigantic bonfire or whether it was tamed by cultivation into the firing of fallowed fields. As early as 1692 Richard Frame put it to doggerel:

Then with the Ax, with Might and Strength,
The trees so thick and strong
Yet on each side such strokes at length,
We laid them all along.

So when the Trees, that grew so high
Were fallen to the ground,
Which we with Fire, most furiously
To ashes did Confound.

In contrast, the long hunters burned the woods as they were. In 1803, on a tour across the Allegheny Mountains, Thaddeus Harris commented "with regret and indignation, the wanton destruction of these noble forests. For more than fifty miles, to the west and north, the mountains were burning. This is done by the hunters, who set fire to dry leaves and decayed fallen timber in the vallies, in order to thin the undergrowth, that they may traverse the woods with more ease in pursuit of game." Some settlers even gathered potash from deep burns to sell on the market, as a later generation collected bison bones on the plains to grind into fertilizer. Settlers fired barrens and woodlands to promote pasture. Those who sought blueberries and huckleberries burned on a cycle of perhaps three years (a practice that continued in places through the 1950s). And of course fires of all causes escaped. Yet the burning persisted as a low-grade, background-count disturbance.[9]

To be transformative fire needed the leverage provided by the axe, or by the supplemental burning from fossil fuels. The latter heralded a new era, congealed in the mid-19th century, that made Pittsburgh the urban complement to Philly. Philadelphia bonded an Atlantic world to a wooded land; wood built its cities and its ships, and powered its hearths. Pittsburgh bonded coastal colonies to the continental interior, and eventually to a civilization powered by fossil fuels.

⸻

The wave of logging moved from Maine south, and from the coast inland by rivers. In Pennsylvania the Susquehanna River opened the interior to the axe. The white pine went, often in log rafts down the river. The cut was serious, but oxen, hand-hefted axes, and seasonally deep streams could only move so much.

The shift came after the Civil War when rails replaced rivers, and coal, cords of fuelwood and rafts of charcoal. Industrial combustion pushed logging to new plateaus, both literally and figuratively. Rails punched over ridge after ridge, pushed through valley after valley, cleaning out prime timber and leaving behind slash and brush. When George Innes painted his famous *Lackawanna Valley* in 1855, the locomotive looked no more sinister than a flock of sheep, and the roundhouse outside Scranton resembled a big barn in what remained a rural, mostly pastoral setting. Twenty years later no one would call the landscape that rail and axe had remade bucolic.

By the 1870s the Big Cut was on, and continued for 40 to 50 years. By the time it wound down Pennsylvania was cut over, as Maine and Massachusetts had been cut over. In his 1882 commentary C. S. Sargent observed that "the pine of New England and New York has already disappeared. Pennsylvania is nearly stripped of her pine, which only a few years ago appeared inexhaustible." The axe didn't stop with white pine. It cut hemlock for its bark, like hunters killing buffalo for their hump, leaving the carcass behind. It took anything big enough to use, if only for charcoal.

And what the axe didn't take wildfire did. The Big Cut sparked an era of big burns. The railroad made both possible. The locomotives that underwrote commercial logging cast embers profligately. Pennsylvania's two fires, the living and the lithic, merged, and like a backfire meeting a fire front, the flames rose mightily.[10]

The clearing of the hill forests and the burning of promiscuous slash upended habitats, which led to widespread flooding and erosion, which led to pervasively wrecked landscapes. Meanwhile agriculturists continued to clear, by now into areas not suitable for farming. And industrial-strength market hunting further stripped the scene. The fabulous wildlife that had sustained early pioneers was devastated. Elk, passenger pigeons, mountain lions, fishers, martens, beavers, and bison were gone; populations of turkeys, grouse, deer, bobcat, otters, and bear plummeted; even songbirds were shot. Fed by slash the fires were so intense that the old woods could not regenerate. What did sprout was sometimes subsequently burned by hunters to open up passage through the thickets.[11]

Not everywhere felt the clash with equal fervor or at the same time; not every place felt the upheaval with the same intensity. But enough did, or were affected indirectly, to make Pennsylvania a national example of what unrestrained capital and fires could do. The chronicle preserved by fire-scarred trees shows a phase change, the pyric transition, and with the conversion to a fossil-fuel society, a radical shift. Burns became more frequent and savage; even northern woods like those along the Pine Creek Valley had their fires triple in number, saw the loss of multidecade fire-free periods, and watched flame divorce from drought.[12]

But there are plenty of eyewitness accounts, even photographs. In his *Report on Forests* for the 1880 Census, Charles Sargent estimated that Pennsylvania burned 685,738 acres, more than any other state north of the Mason-Dixon line (most of the South's fires were part of open-range herding), and second among states in the economic damages caused. He noted that "merchantable pine has now almost disappeared from the state, and the forests of hard wood have been either replaced by a second growth or have been so generally culled of their best trees that comparatively little valuable hard-wood timber now remains."[13]

The perverse fire fugue between industrial and landscape burning had a counterpart in efforts to contain both burns. After the fires came a still small voice of conscience and concern that went under the name of conservation and became a clamor for state intervention to stop the havoc. In 1875 the Pennsylvania Forestry Association formed, the first in the United States, to campaign against the havoc. A forest commission reported in 1895 that the state's wooded heritage had contracted more than 90 percent, below what wood and water supplies and public health needed. Offices for forest and game commissioners followed, along with state forest (and later game) reserves, eventually setting aside over three million acres. Smoke animated the crisis, made it publicly visible, but what united a broad constituency was watershed, and what that meant for climate, floods, and soil loss. It also provided the constitutional means for federal intervention; after the 1911 Weeks Act, a national forest was gazetted in the Alleghenies. The first state forester, Joseph T. Rothrock, became a national voice for forest protection. Pennsylvania's most celebrated forester, Gifford Pinchot, whose ancestral estate in Milford could look both to New York and Pennsylvania, carried the campaign across the country. By the end of the century Penn's woods were on their way to becoming Pinchot's.[14]

─────

From its origins forest conservation identified three tasks: stop the fires, replant trees, and mark the boundaries of protected reserves. It was emergency medicine: first, stabilize, then transport. Restrain, then rebuild. Not only did fire seem the most immediate crisis but the one most easily presented to the public and the least threatening to industry.

The great voices of forestry all argued the case. In 1896, less than a year after becoming commissioner, Joseph Rothrock declared that "there is no subject of equal importance to the Commonwealth which in the past has been so apathetically considered by the mass of our citizens as forest fires." The next year the legislature passed two fire laws that established an apparatus for fighting fires and for penalizing those who set them. A year later Gifford Pinchot, newly installed as chief of the U.S. Division of Forestry, thundered with abolitionist zeal that "like the question of slavery, the question of forest fires may be shelved for some time, at enormous cost in the end, but sooner or later it must be met." He followed in his *Primer on Forestry* with the conclusion that "of all the foes that attack the woodlands of North America, no other is so terrible as fire." Later, as forest commissioner he wrote a guidebook for "Pennsylvania boys" that explained how to control fires, since "we cannot have this wood supply as long as fire endangers every acre of growing trees, and burns about 200,000 acres of forest land every year." George Wirt, long-serving forest commissioner, explained in 1927 what uncontrolled burning meant. "Forest fire is a force which does immediate damage. . . . No forest means no water; no water means no agriculture. Then come floods, drought, pestilence and death."[15]

By the 1920s a political framework of state forests and a basic apparatus to control fires through wardens was in place. The 1930s saw a denser infrastructure created through the Civilian Conservation Corps (CCC), which had 158 camps in Pennsylvania and divided its efforts between tree planting and fire protection. (Many state parks evolved out of CCC camps.) In the postwar era fire protection mechanized, from engines to aircraft, and folded into civil defense. In the 1970s the Bureau of Forestry felt the first tremors of postwar environmentalism, shrugging it off as though it were a swarm of gnats.

Over the years forestry's narrative in Pennsylvania had simplified into the story of its astonishingly successful suppression of fire. Historian Samuel Hays recalls hearing former state forester Jim Nelson tell an audience that "because of the destruction of the forest by fire, the Bureau of Forestry was established." Controlling fire had allowed the native forest to regrow and the planted forests to mature. In truth, the wave of slash-fueled wildfires that had swelled in the late 19th century ebbed by the mid-20th thanks in good part to an aggressive fire warden network. Between 1930 and 1934, before the CCC took hold, the average annual burn had dropped to 153,181 acres. The CCC era pushed that to 54,624 acres. The postwar era saw the plunge continue. From 1960 to 1964 some 30,453 acres burned annually; from 1970 to 1974, 7,992 acres; from 1980 to 1984, 7,873 acres; and from 1990 to 1994, 5,828 acres. And there it has more or less plateaued, between 2,500 and 5,000 acres annually. The warden system shrank from over 4,000 members to some 1,200. Remarkably, fire suppression strengthened even as the Bureau of Forestry assumed responsibility for fire on all state and private land in Pennsylvania not otherwise served by municipal authorities.[16]

As with the Northeast overall, fire, once so prominent and threatening, slid into the status of an endangered process. It disappeared from the region's environmental narrative. If people knew it, they experienced it in a wood-burning fireplace.

The woods grew back, and then they kept growing, restored along with much of the vanished wildlife, and this time they grew beyond their old biotic boundaries. Pennsylvania was again green, famously so.

But its pine forests were mostly gone, its oak forests were poorly regenerating, choked by overtopping maple and birch, its biodiverse barrens were shriveled, its hardwoods were hammered by exotics—chestnut blight, gypsy moths, emerald ash borer, hemlock wooly adelgid, spotted lanterfly; its population of whitetail deer had become pesty; and Lyme disease threatened to reach epidemic proportions. Fire suppression had overshot, quelling the eruption of abusive burning but dropping the population of good fires below replacement value. Equally, industrial combustion was befouling the habitat as the economy sagged, acid rain dripped on the landscape, the traditional climate wobbled, and a new generation of car-enabled, low-density settlement scattered into formerly rural land (by the early 21st century Pennsylvania had the highest wildland-urban interface [WUI] index of any state). Dead trees mingled with dispersed houses. The ills of the Anthropocene had arrived.

That challenge was not unique to Pennsylvania, but the particular terms of engagement were. Unlike New Hampshire it had enough land to make fire's restoration a landscape project; unlike Maine it had biomes ready to burn and seasons long enough to burn in. In broad sketch, Pennsylvania's fire story is New England's, with ligaments binding it to the Ohio Valley and the South, and to that vast oak forest that channeled a backwoods frontier that poured pioneering Americans across the Appalachian Mountains to the prairies of the Great Plains. As the 21st century matured, both forms of burning seemed poised for revival. In 2009 Pennsylvania, led by wildlife enthusiasts, enacted a Prescribed Burning Practices Act and organized a prescribed fire council. At the same time fracking promised to renew fossil fuels. Pennsylvania's fire fugue was entering a new refrain.

WHEN THE RIDGES ROARED

It was unseasonably warm and dry that fall in 1938. When fires broke out in Cameron County, wardens requested assistance from nearby CCC camps. On October 18, Camp 132 dispatched three crews to the Upper Jerry Run fire and 30 men to the Whitehead fire. On the 19th, Camp 86 sent a crew of 20 to the Ellicott Run fire, where they joined 49 civilians from Emporium, and another CCC crew of 19 headed to the Grove Run fire (behind the Sinnemahoning school house). Camp 132 (Hunts Run) dispatched two crews to the Lick Island fire, or what became better known as the Pepper Hill fire. Other crews were assembled from state highway employees and Works Progress Administration (WPA) men from Emporia. They arrived between 4:30 and 5:00 p.m. By then seven CCC enrollees and their foreman were dead or dying.[17]

It wasn't the worst fatality fire endured by the CCC, but it was the worst crew burnover on record east of the Rocky Mountains, and it reminds us that fire control after the Big Cut was a hard, sloppy slog that claimed lives as well as second-growth scrub.

————

The general contours of what happened are known; as usual, the particulars get lost in the thickets. Inevitably, eyewitness accounts conflict. As always, the assignment of culpability is disputed. The compounding of "small screwups," as Norman Maclean put it for a fire a decade later, overwhelms understanding. Missed reports. Missed location. The wrong crew. The wrong foreman. The wrong fire. The wrong preparations. The wrong tactics. The wrong reaction. And, undoubtedly, the wrong timing, by matters of minutes, perhaps seconds. The other crews on the neighboring fires were also tired, indifferently prepared, confusingly led, but survived. One of the Camp 132 crews did not. After the Blackwater fire in Wyoming the year before, the CCC had issued new fire safety guidelines, to be effective upon their receipt by the camp superintendent. Those new guidelines arrived two days after Pepper Hill burned.

The fires, all of them, were unquestionably human caused. The earliest investigators, from the Pennsylvania Motor Police, reported that the "native residents" were emphatic that the fire "began in such a locality that the idea of the fire starting from some natural cause would be idiotic, as the area is completely isolated from inhabited areas and the only persons in the woods at this time of year, prior to the hunting season, are those persons who would have no reason to be there." The persons or person who set them were never identified, although for years afterwards the police and state foresters received rumors about this person or that having conspired to start the fires—hearsay relayed by or about persons invariably intoxicated. State fire warden George Wirt dismissed them as barely reaching the status of gossip.[18]

So there was no motive assigned. The likeliest reasons were grouped under disgruntlement—revenge for some slight, payback by a game offender for his citation, anger over not being hired, resentment over the CCC camps that, to some minds, took work away from locals. Certainly there was, among the natives, the belief that the boys were "too young," ill equipped, and badly led, which is a refracted way of projecting local unease with the program. The natives could have handled the fire better.[19]

There were two crews. One, under Foreman Adolph Kammrath, had 22 men and arrived on the scene at 1:45 p.m. The fire had started at 9:30 a.m. It is likely this was not the fire they were initially sent to, but one they found along the road and assumed was theirs. The second crew, under foreman Gilbert Mohney, had 21 men and arrived at 2:00 p.m. Both had spent the last afternoon and evening on the Upper Jerry Run fire. They had returned to camp at 5:30 a.m. the morning of the 19th. Six hours later they were headed to Pepper Hill Run. A few had probably caught a few hours of sleep, but for most, it was too hot, and they were too unsettled. When the camp fire gong sounded, most hurried to the mess hall for a quick lunch. A few asked to remain—they were too tired to continue. They were ordered to go. All climbed into trucks for what Pete Damico later recalled with no taint of irony as "the greatest adventure of my life," as well as his greatest tragedy.[20]

There was fire at the foot of the ridge and at the top when the crews arrived. Along the road they scraped line for 200 feet through leaves and litter and set a backfire. Kammrath, a certified fire warden, took the "easy" side of the fire; Mohney, not a warden, the "more difficult." Mohney had never worked with this crew on a fire before. William Houpt, camp forester, then arrived to take charge, viewed the scene from the road, ordered more help, then instructed Kammrath and Mohney to take their crews to the ridgetop where the fire head lay, halt its spread, and cut line downhill along its flanks. The more experienced Kammrath refused and stayed with the low fire. Around 3 p.m. squad foreman Mohney and his crew began a long hike up slopes that observers reckoned at 30 to 85 percent through dense second-growth oak. Fire Boss Houpt then left to bring his own crew to the scene. Soon afterwards the wind gusted.[21]

Mohney's crew was strung like beads down the slope. A few were near the summit, most about three-quarters up. Mohney himself was a straggler, struggling to keep pace; it had taken him an hour and a half to ascend half the slope. The lower crew stopped to rest. One boy bit into a sandwich he had hidden in his shirt before suddenly yelling, pointing downslope to a surf of flame, and fainting. Well before Mann Gulch, the boys at Pepper Hill found themselves in a race that couldn't be won. The flames struck. Mohney was the first to fall, which left the crew leaderless. Those near the top scrambled up and over the rim. The group still on the slopes scattered. A few ran up and to the right. Three climbed trees to see if they could find an escape route, then scampered down and followed the voice of Andrew Kiliany, who vocally guided them through the smoke to a safe area. Others sought refuge on a large, flat boulder, though two slid off during the half hour or so it took the flames to pass through. Pete Damico "fell exhausted at the foot of a tree" before later regaining consciousness, alive but burned on his legs and hands, hearing the screams of a "buddy, lying only five feet away from me" and "terribly burned." He could see another crewman, "lying on the ground, his face buried in the earth." With the flames passed, the three walked toward the road, though one, "whose clothing was entirely burned off him, dropped naked and exhausted." The survivors on the rock also wended toward the road. They found two men burned to death. "Boring was on his knees with his hand on his head. The other fellow was also on his knees with a hand and his head resting on a rock." They heard Bogush. Damico "heard him holler, 'Jesus, save me.' But we could do nothing for him. He did not recognize us at all."[22]

The horror rippled out to inadvertent onlookers. William Miller observed the progress of the fire from his farm "directly across from the mountain on which the boys died." About 3:30 p.m. he heard "screaming, yelling, and praying coming from the boys on the mountain," so "harrowing" that Miller, "an aged man," had been unable to sleep at night since the tragedy. In a sentiment echoed by his wife and his daughter-in-law, he claimed that a "race horse could not have fled ahead of the flames." They saw no reason to have sent Mohney's crew to the head of the fire.[23]

Shortly after 5 p.m. the fire finished its brief run. It left five dead on the hill: Gilbert Mohney, Basil Bogush, John Boring, Howard May, Andrew Stefphanic. Ross Hollobauth and Stephen Jacofsky died in hospitals the next day. George Vogel lingered until November 2 in the hospital at Renovo. Two of the boys were 17, two 19, three 18. Mohney was 38.

Before daylight ended rain fell. The fiscal cost stood at $186.64, of which labor accounted for $144.73, vehicle mileage $20.80, and lunches assembled from bread, cheese, mustard, butter, apple butter, pork and beans, canned meat, and dill pickles from Lloyd's Market, $21.01. The final burned area topped out at 67 acres.[24]

———

The tragedy was a sensation locally. Over 600 people attended the memorial mass at St. Mark's Catholic Church in Emporium. Crowds watched trains deliver the bodies to family. Besides the initial reports by the motor police, the Cameron County coroner held an inquest (revealingly labeled an "inquisition"), the district attorney investigated, and the army held a full Board of Inquiry. Few tragic fires have so complete a record. Few were so notorious in their time and so forgotten later.

The CCC was an immense operation. Pennsylvania was a cameo of the country, and like elsewhere those camps installed an infrastructure for fire protection almost overnight. The camps themselves were jointly managed. The U.S. Army ran the camps proper with reserve officers. A "technical" agency, in this case the U.S. Forest Service, through the Pennsylvania Bureau of Forests and Waters, oversaw the actual work projects. Both the reserve officers and the foresters had a vested interest in presenting their actions in a favorable light. The coroner did not. If anything, his jury became the vehicle for channeling community outrage.

Dr. J. D. Johnston, Cameron County coroner, assembled six "good and lawful men," reportedly experienced woodsmen, to review the case. The cause of death was readily determined. The deceased "perished by burns sustained by fire," while engaged in "the performance of duty," none of them drunk or drugged. That took a dozen lines of text. The remaining 3.5 pages judged the contributing causes, and may constitute one of the minor masterpieces in the literature of fatality fires. The stakes were high—not only the loss of so many boys, but the realization of the "need for the future service of C.C.C. enrollees and members of the Technical Service of the U.S. Forestry Department under the supervisions of the U.S. Department of Agriculture in the conservation of our forests." There were lessons to be learned. Johnston's report was a document of contained fury.[25]

Unlike recent reviews the study did not obsess over the physical circumstances of the fire or refuse to match rules with behavior. The fire burned up steep slopes, in flashy fuels, on a hot, dry afternoon, boosted by gusts of wind. Of course it was—for less than an hour until it crested the ridgetop—an intense, fast-moving, flaming front. The issue was what people were doing around it and whether they belonged there. The 10 Standard Fire Fighting Orders were two decades away; the 18 Situations, four. They weren't needed. The jury could peel back from the testimony the layers of liability, and they knew unfitness when they saw it, and they were outraged that seven boys—and they were boys—had died under such circumstances. The lengthy indictment was a kind of punitive-damage levy against the responsible agencies.

It was all about people. Even the fire was "incendiary." Neither the camp commander or the project supervisor had properly organized the camp's safety council "as such Safety relates to forest fire fighting by the enrollees." The new instructions for CCC fire duty had been "promulgated on April 20, 1938," but did not reach Lt. Rodman Haynes, Camp 132 commander, until October 21, an inexcusable "laxity." The camp's enrollees had not been instructed in firefighting despite manuals available since May 1937. The enrollees had not understood their work environment—the Jerry Hill Run and Pepper Hill fires were, for many, their first experience. Still, regulations prescribed that "each crew" would have a "fire warden" as foreman. Mohney's did not. Instead, fire warden Kammrath took the "easy" side of the fire and let the unqualified Mohney take the "more difficult." William Houpt, the camp forester, did not inspect the fire fully, gave orders that put men in the potential path of the fire, and had left the scene during the critical blowup. Gilbert Mohney's "physical make-up disqualified him for forest fire fighting." The upper chain of command, both the camp commander, Lt. Haynes, and the project superintendent, Earl Getz, were absent from camp even though it was still fire season; their delegated replacement, William Getz, had no experience with fire. No one, save the enrollees, escaped blame.

"It is the consensus of opinion," the jury summed up, "that no one individual was more responsible for the death of the above-named deceased more than any other individual." But those listed as contributing "had been lax and negligent in the performance of their duties." The culpability, however unintended, went to the top. "We believe some responsibility also rests upon the superior officers of both the U.S. Army and the Technical Service." All deserved reprimands and penalties.

The army's Board of Inquiry assembled an extraordinary dossier of statements by those involved. It identified Gilbert Mohney as the proximate cause of death, both for his poor judgment in accepting the assignment to leave an "unguarded fire" and ascend the slope and from his "early exhaustion" and death, an "example" that "caused his party to straggle and become disorganized." Other contributing causes were lack of training, indifferent safety committees, and the "failure of both the Company Commander and the Camp Superintendent to take proper measures to comply with Paragraph 'G' page 2 and Paragraph 1, 2, and 3 page 142, Civilian Conservation Corps Safety Regulations." Even so, such neglect "made no contribution to the disaster."[26]

That was too much for the authorities. The Bureau of Forests and Water requested a report, conducted by the U.S. Forest Service, which answered the coroner's charges point by point. It insisted that everyone had acted "in accordance with established State Department [Bureau of Forests and Waters] procedure"; even the order to proceed to the ridgetop and then build line downslope was "in accordance with instructions which have been issued in forest fire publications of this Department, for the last twenty-five years." (As the saying goes, wildland fire is full of bad practices that turn out okay; except this time it didn't.) There was little evidence of "laxity," and if any existed it "had no direct bearing with respect to the death of the enrollees." The one questionable decision belonged with Foreman Kammrath who "did not use the best judgment in attacking the fire," though it was impossible to state whether his actions contributed to the deaths. "We believe not." Nitpicky attention to rules and regulations was neither instructive nor productive. The report concluded that "as far as we can see, no one . . . connected with the CCC activities, can have attached to them, separately or collectively, any blame for the unfortunate trapping of the men on the fire line." Further, it "*must be remembered however, that the results of this particular fire may not have been prevented under any circumstances* [italics added]."[27]

Still, "on the assumption that disciplinary measures cannot be avoided," the inspectors recommended that Superintendent Getz, Forester Houpt, and Foreman Kammrath "be suspended without pay for a maximum period not exceeding thirty days." However much this satisfied the foresters it did not appease the CCC. On March 15, 1939, R. C. Beard of the U.S. Forest Service wrote his counterpart in the Bureau of Forests and Water to explain that Director Robert Fechner of the CCC had "insisted on much more drastic action." The Forest Service had resisted, but "after considerable controversy" between agencies, they had agreed that Forester W. E. Houpt and E. D. Foye, the regional director of CCC camps for the Bureau of Forests and Waters, would receive 30 days leave without pay, Superintendent Earl Getz would be demoted to foreman, and Foreman Adolph Kammrath would be dismissed. The Forest Service and Beard personally felt "that this action is unnecessarily drastic but since the instructions come from the U.S. Secretary of Agriculture, and are approved by Director Fechner, there seems to be no point in making any protests."[28]

———

Perhaps nothing better was possible. In six months Nazi Germany would invade Poland; the CCC was two years away from being dissolved. The postwar era sparked a cascade of crew fatality fires beginning with Mann Gulch and continuing through Loop that washed away the memory of earlier tragedies. Pepper Hill belonged with another era; the American fire community forgot. But the Pennsylvania Bureau of Forestry did not. Nor did enrollees. Some 8,000 contributed pennies to pay for a memorial plaque. Hunts Run Camp 132 erected a stone monument for the plaque to rest on.

THE BURNED-UNDER DISTRICT

On May 27, 1962, the town of Centralia burned its town dump. The project was part of a general cleanup prior to Memorial Day and to encourage townsfolk to use the legal dump rather than the illegal ones scattered around the borough. The fire department did the burn and then doused it. But the flames persisted. Eventually a bulldozer was called in to help mix water with the smoldering garbage and it exposed a cavity leading to a coal mine. The fire had entered a mine shaft. A dump fire was now a mine fire, and the refuse of daily life burned alongside the lithic remains of debris piled in the eponymously named Pennsylvania Era.[29]

The casual, traditional use of fire had interacted with new fuels to create a slow disaster. No one wanted to admit blame, no one wanted to accept responsibility. What could have been stamped out relatively easily at first became more complicated and costly the longer it went on, and on, each delay further ratcheting up difficulties and expenses. Mining in the extensive labyrinth of tunnels ceased. Early on authorities recognized that the cost of extinguishing the fire was orders of magnitude greater than the commercial value of the town. The city tried to hand the problem to the Lehigh Valley Coal Company, which owned mining rights but demurred; it then went to the state, which handed it to the feds, who handed it around possible agencies, none of whom wanted the job. Meanwhile the fire went deeper and farther until the prospect of containing it might itself require drafts of geologic time.

Each of the too-little, too-late attempts to contain the spreading coal fire failed. The fire burned under the town of Centralia, and then under Byrnesville. Sinkholes developed over collapsing shafts; steam and noxious gases rose through holes and percolated through fractured rock upward into basements and neighborhoods. The emissions became a matter of public health, and the holes, of public safety. Residents suffered from carbon monoxide poisoning and oxygen deprivation. In 1981 a 12-year-old boy nearly died when a sinkhole opened up beneath him. In 1984, Congress expended $42 million to relocate the townsfolk of both Centralia and Byrnesville, save for a handful who chose to stay. As residents left, the state invoked eminent domain to seize their property and raze structures in what seemed an effort to wipe away the very memory of the fiasco.

Centralia reverted to a feral landscape, earning status as America's most famous coal mine fire, and inevitably a tourist attraction.

What began as anecdote evolved into apologue, and in time into allegory. Irony came simply, even smugly. Centralia became a smoldering landfill of symbols, some of which entered into deeper shafts of inquiry. But it has not been seen as a concise cameo of fire history, the gist of which is that the Centralia fire jostles together, as few places can, the two grand realms of earthly combustion.

Pennsylvania is the hearth of fossil-fuel fire for America. Anthracite coal was noted in the Ridge and Valley province by the late 18th century. Bituminous coal was found later in the Allegheny Plateau. In 1859 the first successful oil well was drilled at Titusville. The earliest mechanisms for transporting coal and oil—canals, pipelines, rails—emerged to connect the sources of fossil fuel to their industrial and urban sinks. An electric grid followed.[30]

The process fed on itself. The earliest use of steam engines, from Newcomen's to Watt's, was to pump water from coal mines. Engines to power transportation by boat and locomotive made possible still more production. As much fuel as the larger project burned, it exposed still more. It also began to leverage surface fire. By making possible Pennsylvania's Big Cut it stimulated a wave of big burns. Industrial combustion was supposed to be confined to closed combustion chambers and to replace open flame; but its indirect consequences loosed free-burning fire across the countryside. Bad fire began to crowd out good fire.

The mines (and wells) left an oft-trashed landscape. The slag, or culm, littered the region, a perverse landfill of industrial combustion. Chocked with coal many culm banks burned, smoldering counterparts to dump burning. A fair fraction of mines, too, burned. At the time of the Centralia fire, there were 11 in the bituminous region and 16 in the anthracite. All of them were deemed more threatening than the distraction at Centralia. Mine and culm bank fires were the escape fires of industrialization, the equivalent to the field fire that spilled over into the woods. Pennsylvania was the Burned-Under District of industrial America.

But there were coal mine fires wherever there were mines. They burned in West Virginia and Colorado. They burned in every continent, save Antarctica. Coal seam fires in Borneo have been dated back 13,000 years. The coal beds themselves contain the fossil char (fusain) from fires in the geologic past. The Pennsylvania coals are especially rich. An estimated 10 to 20 percent of the coal from the Carboniferous Period is fusain.

What happened at Centralia makes a useful exegesis on this process because it demonstrates, in material ways, the intersection between the two fires. The hills and valleys were honeycombed with mine shafts that tracked the warped beds of coal. When anthracite yielded to fuel oil and bituminous coal, many mines were abandoned, at which point rogue mining often gnawed at the pillars that had supported the mines, encouraging collapse. Where the seams came near the surface, they were strip-mined, which frequently cut across the underground shafts. The landfill at Centralia grew out of just such a strip mine, and the dump fire burned into one of the shafts, which should have been plugged with incombustibles before the landfill was opened but was not. Burning landfills was, by state law, illegal, but was cheap, common, and traditional. A natural gas pipeline running near town also crossed the path of the subterranean burn.

The burning caused the geologic foundations of the town to collapse. But the greater threat was slower and unseen. The emissions that vented or leached upward were toxic, not only with carbon monoxide and carbon dioxide but with metals like mercury and poisonous compounds forged in the combustion chamber that was the mine; unlike commercial coal, it was not sorted, and unlike factory furnaces, there were no chimneys, much less scrubbers. This was a deeper hazard but one not immediately visible and one not handily controlled. The only serious solution was to end the burning.

The story might stand for that larger fire regimen created by burning coal whose emissions threaten to unhinge the Earth's climate and make former habitats unusable, but are not easily contained because control can only succeed by extinguishing the source combustion. That burning won't cease anytime soon. We can expect some dramatic relocations, along with nasty social and political quarrels over who is responsible and who should pay and why the problem wasn't tackled early when it could have been addressed with relative ease and modest expense. Instead, each passing year only embeds the issues more intractably until the problem becomes everyone's and no one's, and it catalyzes a new era of bad burns and what people have begun to call the Anthropocene is renamed the Pyrocene.

THE FOREST IS YOUR FRIEND,
THE FOREST FIRE YOUR ENEMY

The forest is your friend . . .
Our forests are almost gone.
They will grow again if fires are kept out.
The forest fire is your enemy.
— GIFFORD PINCHOT, *FOREST PROTECTION: FIRE*
PREVENTION AND EXTINCTION (1922)

So tightly bonded are forestry and fire protection that it's easy to forget how uneasy was their formative partnership. Forestry emerged as a graft on the rootstock of European agronomy: it meant silviculture, the growing of trees as a crop. Its alliance with the state emerged when tree plantations were used to reclaim sandy wastelands around the Baltic and the Landes or to heal eroded hillsides. Fire was a nuisance like bark beetles or leaf blight, and something that resulted from bad behavior by people, not a fundamental feature of a forest's ecosystem. Central Europe, after all, had no natural fires. America's first professional forester, Bernhard Fernow, an émigré from Prussia, did not even regard fire as part of forestry's mandate but as a precondition for its practice. Forestry schools did not include fire in their curriculum. As late as 1953, A. A. Brown, then director of fire research for the U.S. Forest Service, lamented that American forestry training, "influenced strongly by European traditions," was wildly out of sync with practical needs. "The effect of this has been that many young foresters have found themselves on a job of which four-fifths was protection of the forest from fire, but with their training in inverse ratio."[31]

Such attitudes could not long survive. Free-burning fires were everywhere on the American scene. They could not be dismissed, as Fernow did, as simply the expression of "bad habits and loose morals." Along with reserving lands from the axe, fire control was the foundational charge to administrators. Writing in 1910 Henry Graves, chief forester of the U.S. Forest Service, reckoned that fire protection was 90 percent of American forestry. In his classic 1914 study of fire in California, Coert duBois declared that "American foresters have found that they have a unique fire problem, and they can get little help in solving it from European foresters. . . ." Systematic fire protection would be America's contribution to global forestry. Firefighting became the romance of American forestry. Paradoxically, if wildfire was the principle (or at least most visible) threat to forests, it was also the most profound lever by which forestry became powerful. Over time fire control so dominated agendas that forestry bureaus became, in the public eye, and often in public law, fire services. Even when the fire challenge morphed from fire's removal to its restoration, fire remained under the auspices of forestry bureaus, and so under the lingering heritage of classic silviculture. It's sobering to read 19th-century forestry tracts that prescribe, as a first step, to clear away the wild woods so the land could be cultivated with orchards and plantations. So, too, forestry's fire protection mandate persisted long after the great burns had passed and alarm over a prospective timber famine had evolved into a genuine fire famine. Of all the disciplines that might have laid claim to fire as a subject for research and management early foresters were probably among the most ill equipped.

———

Pennsylvania was both blessed and cursed as a first mover in American forest conservation. For this it can point to the Big Cut and the Long Burn that fed on its wastes. The state's reforms moved in loose sync with those at a national level. In Samuel Hays's words, it constituted "one of the nation's most celebrated cases of state forest management." In 1875 it organized the oldest state-based forestry association in the country. Not having large expanses of still-public domain, it had to reclaim one by purchase. In 1895 it established a state forest system, eventually totaling over two million acres, later supplemented by state parks and game reserves. Through a succession of public foresters, it helped define the national agenda.[32]

They were a striking gallery—oversized characters—of prophets and patriarchs. They were powerful voices, charismatic personalities, and often long-serving presences. Joseph Rothrock led a chorus for reform that fed not only into Penn's woods but the nation's. George H. Wirt, chief fire warden, directed the Bureau of Forests and Waters fire operations from 1915 to 1946, including the CCC era that installed an infrastructure, and made Pennsylvania a standard in state forestry. And of course there was Gifford Pinchot, the most famous forester of his generation.

The Pinchot family estate, Grey Towers, at Milford looks north toward the Adirondacks, south into New Jersey (and the pine barrens that Pinchot later studied), and west into the Pennsylvania of the Big Cut and Long Burn. He studied briefly under Dietrich Brandis at Nancy, France, absorbing the curriculum given to British imperial foresters, joined the National Academy of Sciences (NAS) Committee on Forests, helped draft the 1897 Organic Act for the forest reserves, succeeded Fernow as director of the U.S. Division of Forestry, became first chief forester of the U.S. Forest Service in 1905, until fired by President Taft for insubordination in 1910, then campaigned for conservation nationally before becoming Pennsylvania's forest commissioner from 1920 to 1922 and then twice state governor.

Throughout, he recognized fire both as a genuine threat and as an issue on which to rally public enthusiasms for conservation. The National Academy of Sciences Committee on Forests identified fire protection as one of three fundamental questions. He began his directorship of the Division of Forestry by investigating the fire scene, with the idea of showing how much fires cost the country. In his *Primer on Forestry* he identified fire as the worst foe of American forests. He followed with a study of the pine barrens in which he, improbably, equated the crusade against fires with that against slavery. As chief forester, he wrote in the *Use Book* that fire was the first duty of rangers. As forest commissioner for Pennsylvania he wrote a tract on fire protection that effectively argued that the most direct way to recover the state's vanishing forests was to spare them from fire. Firefighting was both good forestry and good public relations.

In 1947 he published his memoir, *Breaking New Ground*, part of American forestry's holy writ. He noted happily that he had been "a Governor, every now and then, but I am a forester all the time—have been, and shall be, all my working life." In 1960 the family donated Grey Towers to the USFS. In 1963 it was listed as a National Historic Site, and later a National Historic Landmark and Pennsylvania state historical marker. A Pinchot Institute continues the legacy by promoting conservation. By then Grey Towers had assumed the status of a secular shrine.[33]

All in all, it's an extraordinary legacy, felt with particular force in Pennsylvania.

In the late 19th century state-sponsored forestry was a start-up, and it benefited from exceptional founders who could meld vision with practical politics. But organizations mature. Landscapes, once shielded from abuse, also mature. Organizations with charismatic founders can hold to their heritage longer than may be useful; they may become deferential, conservative in unthinking ways, and defensive if their identity is staffed by a professional society of similarly shared values. They retell the founding sagas and creation story. They often look to a past designed for challenges that no longer exist. Powerful, inspirational founders are a blessing, until they become a curse.

Political forestry wasn't unique: think of the FBI still shedding the toxic legacy of J. Edgar Hoover or the U.S. Geological Survey (USGS) wrestling with an institutionalized hagiography of John Wesley Powell. Forestry had an additional burden of pride in its status as self-proclaimed profession. Foresters were the experts on all matters pertaining to woods, and later, wildland fire. In truth they became very skilled at fire suppression, which fit easily into silviculture; they proved poor at adapting to fire restoration for general ecological enhancement. The torch passed mostly to wildlife enthusiasts, people attracted to wilderness, or partisans of cultural landscapes like tallgrass prairie and longleaf pine.

Here forestry's stiff-necked appeal to technical knowledge and professional authority worked against it. And that pointed to another quirk. Forestry's founders often had minimal training in academic forestry, much of which had little to do with the American scene anyway, but they were broadly educated, politically astute, articulate, and culturally attuned. Their successor generations became more technically competent, keenly schooled in forestry, but too often culturally isolated, unable to engage the public. They fell back on old tropes, relied on authority, and appealed to firefighting as a heroic task.

So American forestry has struggled to overcome its founder effect and the era of political power it enjoyed. Places with a strong tradition had the tougher struggle. Since so much of that heritage had involved fire protection, transforming fire control into fire management could mean turning the creation story inside out. Measured against its founding threat forestry had succeeded; and that ordeal had bequeathed a kind of heroic age.

Now the need was to reinstate fire, and the story grated and splintered as it moved cross-grained to the image of the agency and the mission it had so long proclaimed to the public.

That happened nationally as a generation that watched the Big Cut and Long Burn sought to shield public lands in the West from similar havoc. And it seems to characterize Pennsylvania. Foresters stumbled to move from Old to New Forestry, to find some way to honor the past without being shackled to it, to engage with fire on new terms. The torch passed to nongovernmental organizations (NGOs) and the game commissions that focused on habitats critical to rare species and game, places not amenable to commercial logging. Foresters became partners, not leaders. They struggled to move out of the fire-cast shadow of Grey Towers. Penn's woods had become Pinchot's, and Pinchot's woods passed out of forestry.

REBURNS

In ways no one would have predicted even a decade ago, fire is reasserting its role in Pennsylvania. In colonial times, the state bridged north and south; now, it has the potential to span northern biomes and southern fire traditions. It is reacquiring patches of landscape burning and revitalizing its stock of industrial combustion. Prescribed fire and fracking are renewing Pennsylvania's pyric fugue.

The pressure for prescribed fire came mostly from wildlife enthusiasts. Pennsylvania forestry was perhaps too much a victim of its own success in taming the industrial strength slash and burn of the Big Cut and its aftermath. It had waxed great on its fire protection mission. It now found it hard to shed over a century's dedication to reducing fire, a project in that, by almost every metric of the day, it had succeeded. With fire in hand, foresters assumed they could address their core identity: timber. Penn's woods were supposed to be green, not red or black.

But those woods were less robust than they should have been. Trees had returned, not the old forest and other landscapes that had made

Pennsylvania famous for wildlife. The conviction gnawed that those half-restored landscapes needed fire. Barrens needed fire to sustain berries and biodiversity; woodlands needed fire to keep the land open, dappled with sunlight, and lush with shrubs and grasses. Oaks and pine, in different ways, needed fire to prevent shade-hoarding hardwoods like maple and paper birch from overtopping them. The surface needed to be more than windfall and hard-packed leaves. In particular, the Pennsylvania Game Commission decided it needed different fire regimes on its lands than the Bureau of Forestry had on its.

In 2005 it hosted an omnium gatherum of folk interested, for various causes, in prescribed fire. The Game Commission, the Nature Conservancy, and the Bureau of Forestry agreed to conduct a "prescribed fire survey" to canvass needs and interests through the state. At a gathering in Williamsport in February 2007 the leading partners decided to form a Pennsylvania Prescribed Fire Council. Now that they had something to rally around, prescribed fire enthusiasts—federal, state, and private—came out of the woods. Some 38 organizations, it seemed, used or wanted to use prescribed fire. By now that included, if cautiously, the Bureau of Forestry.

Often those who come later to an idea can speed their progress by relying on the spade work of others. In this case the Coalition of Prescribed Fire Councils had pioneered the basics of a model law, publicity campaigns, and liability requirements, and organizations like the Nature Conservancy, the National Park Service, and Fish and Wildlife Service had stocked shelves with lessons learned. Survey results were disseminated. The council set as its mission "to promote the exchange of information, techniques, and experiences of the Pennsylvania prescribed fire community, and to promote public understanding of the importance and benefits of prescribed fire."

A year later, February 2008, the council hosted the first prescribed fire council in Pennsylvania. Some 300 people attended; with money from the conference and assorted grants, the council developed a website, established working groups, sponsored training, and pressed for legal recognition (and legitimacy) for prescribed burning. The first bill was introduced to the Pennsylvania legislature in August 2008. After amendments, the Prescribed Burning Practices Act passed, unanimously, in July 2009. The Game Commission and TNC conducted their inaugural burn outside

State College in April 2010. By the end of 2015 some 10 agencies had conducted burns at 372 sites. Prescribed fire was a torch whose time had come.

═══════════════

In eerie counterpoint Pennsylvania was also reigniting its heritage of fossil fuel combustion. New technologies of fracking liberated natural gas in the Utica and especially the Marcellus shales. By the time prescribed burners were gathering for their first assembly, the fracking boom was on. By 2015 Pennsylvania may have had 372 sites for surface fire but over 7,700 drill sites for natural gas. Even that 20-fold proportion underestimated the leveraged power of industrial combustion. Coal was declining—with or without environment restrictions, it was too crude to compete. But natural gas was exploding. Both of Pennsylvania's two realms of fire were poised to expand.

The issue for the future is how they will interact. Reintroducing surface fire will cost money but enhance the environment. Redrilling for fossil fuels will bring in money but, directly and indirectly, harm the environment. In the 19th century the damage was felt by watersheds—now, by subsurface waters. It's a dilemma the country overall faces, but only a handful of states has distilled the drama within their own borders. Once again Pennsylvania finds itself in the vanguard of the nation's fire history.

BOG AND BURN

The New Jersey Pinelands

T
HE WILD CRANBERRY IS a wetland species, found naturally along creeks and the edges of bogs. Growing it on the commercial scale practiced by Lee Brothers Inc. requires field irrigation on the model of wet rice cultivation. The plots are flooded over the winter to protect the vines and again during harvest, as berries float and are swooped along and scooped up. The system runs on water. In the New Jersey Pinelands the rule of thumb is 10 acres of watershed for each acre of cranberries. Those watershed acres are forested with one of the most combustible biotas in North America. Bogs abut burns. The regimen of managed water has its parallel in a regimen of managed fire.[1]

Cranberry cultivation is the Pinelands in cameo. Bog and burn are only the simplest of the region's antitheses. But what defines the scene is not just the starkness of its contrasts but their intensity. The water seasonally loosed or sprayed onto bogs is a trivial surface expression of the Kirkwood-Cohansey aquifer—at 17.7 trillion gallons, among the nation's largest. The surface litter and shrubs that carry fire across the forest floor are a fraction of a fuel reservoir that not only replenishes but grows deeper with each passing year. The visual wind sheer of cranberry bog adjacent to burning woods is a tiny tile from a regional mosaic of land uses that places shopping malls next to feral pitch pine, blueberry row crops against Atlantic white cedar wetland and upland mixed oak, and reserved wildlands against conurbation. A 1.1 million acre reserve, of which 660,629 acres are forested uplands and wetlands, lies within a few hours' drive of

35 million people; extend that range to Boston and back from the coast, and add another 10 million. The most densely populated of the American states has more than a third of its landed estate in nature protection; 22 percent of New Jersey lies in the Pinelands reserve. There are few gradations: developed suburb brushes cheek-to-jowl against nature reserve.[2] What really astonishes is that the Pinelands are among the most flammable landscapes in America. Their recorded fire history dates from the earliest European contact. With the rougher contact of settlement came rougher fires. Each new wave of exploitation slashed and reburned the Pinelands, each pass seemingly selecting for greater pyrophilia. The extensive urban development that surrounds the contemporary woods lies next to the biotic equivalent of a munitions depot or an abandoned tenement rotting into combustibles. The Pinelands are perhaps the most famous unknown firescape in America.[3]

———————

The earliest explorers to New Jersey smelled its smoke before they saw its shore, and they viewed its smoke plumes before they knew what those fires were combusting. The land burned. A century of modern records identifies, on average, one lightning-kindled fire a year; some years have none, some a handful. During droughty summers that spark may have lingered in the peat of drained bogs for weeks, sending out tendrils of flame from time to time. But not long after the Pleistocene ice departed, there were people, who undoubtedly burned, and kindled far more starts than nature. Torch and lightning together put fire onto what evolved into the Pinelands. Human ignitions have overwhelmed natural sources ever since.

Either ignition could only propagate if environmental conditions allowed: neither can force fire through a landscape covered in snow or swampy or flush with summer growth. Resting atop an immense aquifer whose surface is checkered with bogs, creeks, and surface seeping, and bordered by floodplains, deltas, and salt marshes, the early landscape was a complex matrix of wet and dry sites—wetlands still comprise some 35 percent of the surface geography. Fires could be set unceasingly, but whether or not they spread depended on the capacity of the land to carry them. For more fire, or a different regimen of burning, people would have to manipulate fuels as they did ignition.

By the time of European contact, the Leni-Lenape practiced swidden cultivation in the floodplains and on patches of better soil, and likely used the upland forests for hunting and foraging. While few explicit records exist of their fire practices—an observer wrote in 1765 that they regularly burned the countryside to assist hunting—analogies with comparable economies throughout the world suggest they burned routinely along regular routes of travel and on sites used for seasonal hunting and gathering. In dry years or when high winds blustered, those confined lines of fire and fields of fire could bolt across the landscape. And if history is any guide, fire littering would have been common. There were almost always sparks on the land; and when that scene was ready, they would fly with the wind. The outcome would have been a dappled landscape of wet and dry patches, with the dry patches burned frequently and the wet ones occasionally slow-combusted, scouring out the basins and keeping them from filling with peat. The resulting landscape was probably one of "small scattered pines and oaks, low shrubs such as blueberries and huckleberries, and some sedges, legumes, and other herbaceous plants." In brief, it resembled hundreds of presettlement ecotones maintained "by frequent and relatively light fires."[4]

That dynamic changed when European settlement introduced felling axes, cereal grains, livestock, swidden not restricted to floodplains, and new markets, all of which allowed the colonists to convert forests into fuels. Settlers could now shift the times and places for fire—its patches and pulses; they could change the reasons for burning; they could add fire as a means to hunt deer, trap muskrat, and promote waterfowl and use fire to promote pasture, from pine savannas to salt marsh grasses burned and harvested for hay. The landscape commenced its chronicle as an ecological parchment inscribed and erased over and again as old practices persisted and new ones were added.

What made this region of the Atlantic shore special was that the point of contact, the posts of New Sweden along the Delaware River, put Finns accustomed to first-contact clearing around the boreal forests of Scandinavia into proximity with the Leni-Lenape. Out of their exchanges came a peculiar hybrid, a fire fusion, that evolved into the exemplar of backwoods American pioneering. Its distinctive material culture included such traits as the rude log cabin and worm fence; its economy, practices like the long hunt; and its fire culture, an alloy of new-land slash-and-burn cultivation with broadcast burning for hunting, foraging, and herding. The model, like the frontier, moved west, but some surely slopped eastward into the Pinelands.[5]

It was, however, a frontier fire regimen, not one for sedentary settlement. The lumpy lobe that became south New Jersey defied the rooted landscapes favored by the English, Dutch, and Germans. Too many of its soils were sandy and starved of organics. In places a lustier loam could support traditional European crops, or the rotation favored by the agricultural revolution, and so the villages that created them. But mostly the land resisted. It became, in the common parlance of the time, a *barrens*. The best one could hope for agronomically in the interior was to domesticate and propagate the indigenous fruits—huckleberry, blueberry, and cranberry. What the barrens had was water, bog iron, and wood, mostly pine and oak in various mixtures. The wetlands had Atlantic white cedar. Along the margins, and in pockets of better soils, flourished such hardwoods as red maple, black gum, and hickory.

Those woods were cleared as rapidly as the technologies of transportation made possible. The axe cut a wide swath first along rivers, then it rode the rails, and more recently it has driven over paved roads. For that first era the forest was felled for sawtimber and fuel. The timber went to shipbuilding, local construction, and export. The firewood fed virtually every industry. Powered by pitch and shortleaf pine, it distilled sap and pitch into tar and turpentine; combined with bog iron, it powered forges and furnaces; combined with sand, it sustained a major glass industry. Perhaps the most interesting fire technology was the thrice-burned charcoal. It was said that the woods would be burned to discourage other uses of the trees, then hewn bolts would be stacked and slow-cooked in special hives to leach away the volatiles and leave blocks of char that could burn by glowing combustion. There was no effort to conserve the woods. When one area was exhausted, another would be felled and burned. The industries were migratory, or were at least migratory in their demand for fuel.[6]

The process quickened when railroads punched into the Pinelands. Its woods furnished ties and fed the engines, which opened up yet more forest to exploitation and scattered sparks like iron from a spinning whetstone. Logging shattered the structure of the forest, and promiscuous burning broke the rhythms of its processes. The rails became the Pineland's new lines of fire, replacing colliers as primary ignition sources. The legacy forest was no longer adapted to the kinds of fires lavished upon it. What had been a barrens, biologically lush but largely impermeable to sedentary agriculture, save for cranberry bogs and blueberry fields, increasingly became a wasteland, fit only for fire or a fire-catalyzed folk economy. (Even the harvesting of pine cones as ornamentals relied on kilns to open the oft-closed cones.)

By the latter half of the 19th century written accounts speak of widespread burning and occasional conflagrations. In his 1878 *Report Upon Forestry* Franklin Hough observed that the region, "comprising a million or more of acres," had been "stripped of wood for charcoal" and "repeatedly been the scene of destructive fires, increasing within the past few years in extent of damage." An 1866 fire burned 10,000 acres. From 1870 to 1871 "nearly the whole wooded portion of Bass Township, Burlington County, was burnt over." Two fires in Ocean County burned over 30,000 acres. The next year 15 to 20 square miles went up in flames. Hough estimated that the "whole country is overrun about once in 20 years by fire." Understandably, wood stocks had plummeted, organic soils had vaporized, shipbuilding had collapsed, and "while nearly nine-tenths of the surface is wooded," residents were "obliged to import nearly all the lumber required for use." In 1894 a single wildfire burned 125,000 acres. A 1915 fire burned 102,000 acres. The estimated annual average burned exceeded 100,000 acres a year. The only use left to the burned forests was to reburn them as fuelwood or charcoal. It was as though the central Pinelands had become a collier's oven that residents no longer bothered to cover.[7]

Gradually, the scene calmed. Scientists and foresters issued reports and forecast damnation unless the fires could be contained. No less a personage than Gifford Pinchot, assisted by Henry Graves, after surveying the scene in 1897, concluded that "the condition of the Plains is due wholly to fire," and that "under the influence of repeated burning . . . the complete impoverishment of southern New Jersey is close at hand unless the fires can be stopped." Soon afterwards he became head of the U.S. Bureau of Forestry, which morphed into the U.S. Forest Service when it acquired the nation's forest reserves in 1905. The next year New Jersey created a Forest Fire Service (NJFFS), staffed with a Pinchot protégé, A. G. Gaskill. Success came grudgingly: much of the landscape was not cultivated, had gone feral, and so had the fires. In 1910 the service began erecting fire towers. The 1924 Clarke-McNary Program brought federal assistance. The Forest Fire Service experimented with aircraft in 1927 and with protective burning in 1928. In 1930 eight wildfires rampaged across 172,000 acres. In 1933 with CCC labor and funding from the Roosevelt administration the national and state forest services established the Lebanon Experimental Forest. In 1934 H. J. Lutz published a review of ideas and evidence on the pine barrens for the Yale School of Forestry, not surprisingly concluding that fire was a formative cause. By 1935 the Lebanon Forest was sponsoring experiments into fire. In 1936 Silas Little, along with E. B. Moore, both students of H. H. Chapman at the Yale School of Forestry, began classic investigations into fire regimes and prescribed burning. The CCC and NJFFS were successfully holding the line, both by creating an infrastructure complete with roads, guard stations, and fire towers, and by fighting fires; in May 1936 three enrollees and two NJFFS firefighters died in a fire near Chatsworth.[8]

The fire community soon appreciated that their best counterforce to fire was fire. It was an old folk art, though one that could go wrong. "The means chiefly employed for stopping the progress" of a wildfire was "by backfiring on the line of the roads; those nearest the fire being used first, and if that failed, the next." If the weather favored bad fires, it also easily loosed set fires, so many backburns were lost. Worse, everyone scrambled to save his own property "without regard to the interests of neighbors or the interest of the whole" and once a smoke rose the landscape was soon saturated with burns. By the middle of the 19th century cranberry growers burned around their bogs, and a 1909 law (later declared unconstitutional) required railroads to abate fuels along their rights-of-way, which they typically did by early-season strip burning. In 1928 the NJFFS adapted the practice to reduce hazard around state forests. That was presuppression; for suppression it relied on backfires set from roads and trails. Little's research gave empirical heft and scientific stiffening to the practice, which the Forest Fire Service adopted publicly in 1948. By 1940 average wildfire acreage had dropped from 50,000 acres to 20,000 acres. But because early-season protective burning had ramped up, the amount of fire on the land overall remained high. It had to.[9]

By now Little reckoned, based on his studies, that most of the Pinelands had been cleared and burned over four or five times. That anything still grew is testimony to the jumbled texture of wet lowlands and dry uplands, and to the unfathomable tenacity of the indigenous biota.

Yet however savage the land scalping, after it had ceased, the woods, the shrubs, and the fauna returned. Perhaps the wetlands had served as refugia for many species, while others tempered themselves to the remorseless flames. Except in enclaves where a village rooted or fields were cultivated, no permanent land conversion had been possible. The human population had crashed from a high in 1859.

When the postwar era replaced rails with modern roads, however, that prospect changed. A new wave of settlement by suburb gathered momentum. It paved over the middle of the state, from Newark to Camden, then spilled to each side like a scree field off a high ridge. A crust of development crept down the shore. Then the Atlantic City Expressway between Philadelphia and the coast split the Pinelands as rails had a century before. In 1976 approval for casino-style gambling in Atlantic City created another apex of urbanization and placed the Pinelands within a shrinking triangle of development that threatened to gradually squeeze the life out of the woods. If allowed to proceed, that process would obliterate the Pinelands as a quasi-natural landscape. Its previously indestructible woods would be reduced to little more than decorative landscaping for sidewalks and patios.

Urban sprawl was, for the Pinelands, an existential threat. But then the Pinelands, through their extraordinary capacity to combust, posed an existential threat to sprawl where the two met. On April 20, 1963, a complex of six wildfires ripped through 162,000 to 183,000 acres of the pine barrens, killed seven people, incinerated 458 buildings, and forced thousands of residents to evacuate. The Black Saturday fire came two years after the Bel Air-Brentwood conflagration in Southern California that effectively announced a new avatar of settlement fire, the reincarnation of rural fire into what became clumsily labeled the wildland-urban interface. Burning in Hollywood's back lot, Bel Air-Brentwood became a celebrity event. Chatsworth had the worst fire.[10]

━━━━━

Nothing so distills the essence of the Pinelands as a firescape as its primary denizen, the pitch pine (*Pinus rigida*). Give it good soil and a sunny site, hold competitors in check, keep its fires on the ground, and pitch pine will flourish. It will grow tall, fat, and straight. Likely it had, in the distant past, assumed for the Northeast the character and habits that the longleaf enjoyed in the Southeast and the ponderosa in the West. Historic accounts speak of grassy expanses useful for pasture. What was labeled as pine barrens may have been a glade-like upland pine savanna, stocked with large trees, woven amid lowland bogs. Heath hens abounded until, with overhunting and loss of habitat, they went extinct.

But few pitch pine would have known landscapes not burned often if not hard. Its adaptations to fire are many and ancient. Like most pines it comes with thick bark, it self-prunes its lower branches, and it reseeds nicely into ash. Like a handful of pines it can carry serotinous cones that open when a flash of flame passes through the crowns. Like a few Mexican pines it can refoliate after fire strips branches. But alone it can sprout new trunks from the root collar when the main one has been seared insentient by fire and can sprout epicormically from branches and trunk into fuzzy pockets of green growth from which new leaders will emerge. If repeatedly burned while young, it may shed its taproot and send out a web of lateral roots as scrub oak does. It can thrive amid repeated, even annual burnings; it can survive serial crown fires.[11]

What form it assumes depends on the fire regime under which it lives. If mild, with frequent surface fire, it will reseed with nonserotinous cones. If severe, with consuming crown fires every decade or two, it will sprout from trunk and branch and favor serotiny. It does not, in brief, display a single trait that adapts it to fire but many traits, a suite ready to be released depending on circumstances. Almost certainly it is the most robustly fire-adapted tree in North America. Only the longleaf can approximate its durability. In extreme forms, hammered by a wave train of high-intensity fires, it can become bent, twisted, dwarfish, more like a shrub than a tree. What chamise is to Southern California, pitch pine is to the Northeast. It is the ultimate fire survivor.

And perhaps the Pinelands' principle cipher. No one doubts that fire has figured hugely in the Pinelands' history, or that, so long as the land remains populated with its original flora, fire is both inevitable and necessary. The issue is what that fire means and how to manage it. Any such contemplation leads to the pine plains, a 15,000-acre parcel of the Pinelands that stands to the complexity of the biota as the pitch pine does to its flora. High-intensity fires sweep over a patch roughly every 8 to 10 years, abrading and pummeling the biota into a dwarfish tangle of pitch pine, scrub oak, and collateral pyrophytes, so selecting for fire traits that seeds from plains pines will assume plains habits even when planted in loam. The plains concentrate the ecology, concepts, practice, politics, and narrative of Pinelands fire.[12]

They invite two perspectives. One views it as a hearth, where fire is purified into dominance and other pressures on the ecosystem have shrunken. It sees fire as a core process, as informing for the surface biology as the Kirkwood-Cohansey aquifer is to the subsurface geology. Outside the plains, fire's presence fractures as flames filter through wet and dry patches and the sieves of less fire-hardy species. The plains are the Pinelands' fire purified. They are the fire counterpart to formerly vast wetlands, where the aquifer seeps through the surface and floods.

The second perspective sees the pine plains as the dregs of all that make the Pinelands peculiar, as though its extraordinary fire history had boiled down the most grotesque features of the regional scene into a pithy distillate. The plains are the 2 percent of what remain after stripping away the specifics, the variety, and the ecological delicacy of the Pinelands. Even fire sheds its subtleties: it becomes singular and monstrous. It no longer wends its way through bog and woods, poking and probing, seizing combustibles and shunning quagmires, killing young oak and gum and promoting pine, but teeters on a knife-edge of blowing up.

New Jersey consists of two crudely equivalent lobes, a northern one drawn by political decisions and a southern one defined by natural processes. The northern belongs with the continental land mass, grading into the foothills of the Appalachians. The southern aligns with other islands and peninsulas that jut into or drop down to the Atlantic. Its ecology depends on how deep its waters and how frequent its fire. That's the same formula of bog and burn that characterizes Florida. From a pyrogeographic perspective New Jersey is a scrunched up, chilled down Florida. The pine plains are to its peninsula what the Everglades and Big Cypress are to Florida's. For each the choice is not whether to have fire or not, but what kind of fire will come. For interior Florida, long given to open-range ranching, the historic solution was to deliberately burn—twice a year, as the adage went. For the Pinelands, violently scalped every century, the fires came as conflagrations every decade or two.

Even amid the Gilded Age the Pinelands fire scene was outrageous and denounced. Unlike western fires, its plumes could be seen from new high-rises in Philadelphia and New York. Incendiarism was commonplace. Odd as it might seem to modern experience, the national fire crisis resided in the Northeast and around the Great Lakes. North Woods Michigan, upstate New York, Maine, and the New Jersey Pinelands inscribed the zone of catastrophic fire.

In 1906 New Jersey established a Forest Fire Service (and Forest, Park, and Reservation Commission) to bring some degree of protection, if not order, to the countryside. The New Deal used the Works Progress Administration and CCC to erect an infrastructure of fire roads, camps, and lookout towers; the CCC supplied crews; and the U.S. Forest Service funded research. The evolved solution pointed to a mixture of protective burning and aggressive suppression. The land, once again, began to recover. Open burning declined, tamed fire replaced feral, and controlled burning for hazard reduction substituted for laissez-faire arson. Then internal combustion began redefining the dominant fire regime. It laid down an asphalt exoskeleton that thickened inward. It gnawed at the interior Pinelands like glowing combustion through a drained bog. Something needed to contain it as the NJFFS had free-burning flames.

The outcome was the Pinelands Protection Act. In 1978 Congress authorized the Pinelands National Reserve, a gerrymandered region of 1.1 million acres that encompassed most of southern New Jersey and was nominally placed under the National Park Service. The idea of a "reserve" was novel: it was not a park, nor a recreation area, nor a preserve like Big Cypress (enacted in 1974). It more resembled the model of a biosphere reserve, which the Pinelands became in 1983 (and internationally, in 1988). In 1979 New Jersey authorized the Pinelands Protection Act, which established a Pinelands Commission, which subsequently led to a comprehensive management plan over 938,000 of the reserve's acres to stymie sprawl from consuming the Pinelands.

One of the few tests on Benton MacKaye's vision of a regional planning authority capable of coping with multiple use and scores of jurisdictions, the national legislation includes 1.1 million acres across seven counties, 56 municipalities, and a handful of federal installations; and the state legislation, 938,000 acres, 53 municipalities, and seven counties. Two military bases, a bombing range, seven large state forests, two major coastal wildlife refuges, two wild and scenic rivers, a national estuary research reserve, a legal wilderness, a handful of endemic species, a rash of entries under the National Register of Historic Places, an international airport, and 312,000 people organized into historic villages, retirement enclaves, and mall-centered exurbs. Today, the Pinelands produce most of the state's blueberry and cranberry crop, protect 43 threatened or endangered animals and 92 plants, and oversee 245,000 acres of forest. Two-thirds of the Pinelands are privately owned.

Under the comprehensive management plan the land use that existed at the time of the Pinelands Protection Act could continue: old houses could be rebuilt on existing sites but new ones could not be erected elsewhere; farmers could still farm but not sell to subdividers; already developed areas could redevelop further and fill in but not expand; forestry could work over woods but not clear-cut into new territory. The upshot is that the historical dappling of wet and dry sites has expanded to include developed and undeveloped. Each has a complex texture. The wildlands have bogs and rises; pitch, shortleaf, Virginia, and loblolly pines; scarlet, chestnut, black, white, and post oaks; blueberry, huckleberry, sphagnum moss, greenbriar, warm-season grasses; hardwoods like hickory, red maple, and gum. Wind, drought, flood, gypsy moth, bark beetle, and fire churn them into various compositions. Only in a few places do pressures push toward dominance by a single species or process. So, too, with the developed sites. They have shopping malls, golf courses, sand mines, trailer courts, Walmarts, churches, billboards, cemeteries, business parks, highway strip malls, retirement communities, farm houses, convenience stores and gas stations, banks, garages, fast food franchises, sewage treatment plants, and nurseries. But while each realm can rework its parts, neither will drive out the other.

The plan has worked. It has survived court challenges, economic pressures, and political maneuvering. It helped that the main corridor of development lay between New York and Philadelphia to the north, and that the two primary turnpikes lay to either side of the reserve. The Pinelands remained relatively isolated. With that founding legislation the asset stripping that had characterized its former history ceased. The Pinelands Protection Act stopped a ruinous, likely irreversible conversion. For a while internal combustion had propagated as promiscuously as open burning, and threatened to replace one fire realm with another. With the Pinelands Commission, however, controlled burning came to internal combustion as it had earlier through the NJFFS for wildland burning.

But stopping sprawl did not stop wildland fire. There was no effort to control the regrowth of the woods as there was to control the spread and intensity of development. Rather, the land, once again, was abandoned, though in the name of nature protection. Deliberate neglect replaced the indifferent abuses of the past. And, as it has so often, the Pinelands renewed itself to burn. Every year the pressure builds. For four hundred years the land had reconstituted itself with fire as a critical, constant feature. Now that eternal flame was becoming more episodic; surface burning hovered at a tipping point for conflagration. There was less wildfire—the New Jersey Forest Fire Service excelled at aggressive initial attack. The acres blackened by wildfire dropped to a fraction of their level a century before, fewer than 3,000 acres a year.

But there was also less controlled burning. Fire officers were restricted by law to burn only for hazard reduction, and they returned to the same traditional sites to fire off strips and occasional blocks to dampen the volatility of the resident combustibles. In a few places burning expanded somewhat, in part due to interest by private landowners; in most, it receded. Air quality, fear of escapes, general liability—what held prescribed fire back across the country retarded it in the Pinelands. Here prescribed burning was justified because it helped suppression. Yet at 11,000 acres a year, and these the same sites burned over and again, it was not keeping pace with the fuel loads piled up by a surging woods. Silas Little had estimated that, under 1963 conditions, any woods burned less than two years previously would carry fire. In 2018 new legislation improved the opportunities to burn more.[13]

Nor was suppression matching resources with risks. The NJFFS operates on a tradition cultivated for over a century and developed in relative isolation. Firefighting is handed down through generations and across clans. It has its own specially evolved equipment like brush trucks, its own presuppression programs of protective burning, its own tactics of pump-and-roll while crashing through the woods. It fills its drip torches with straight gasoline. It builds much of its equipment, and relies on the federal excess equipment program for many vehicles and most aircraft. It has Huey helicopters handed down as discards from the Maine Forest Service, which originally got them as surplus from the military;

it cannibalizes enough parts from its cache to keep one ship flying. It's a scenario for holding the line. The problem, however, is that the past is only prologue to a worsening future. The Pinelands' capacity to burn is ratcheting up faster than the NJFFS's ability to suppress. It can handle one bad fire, even two, but not a dozen. Still, even a fully modern system will fail during the worst case, and it is the worst case that will likely define the future narrative of Pinelands' fire.

Outside certain growth zones, the only movement of land usage the Pinelands comprehensive management plan allows is from the developed to the undeveloped. Sites can revert from farm to forest, or from house to field, but not the reverse. Similarly, lands can transfer from private to public ownership, but not vice versa. It's a formula to resist industrial encroachment. It's not a formula to manage the land so transferred. What the Pinelands crave is the equivalent of the comprehensive management plan for its wildlands. It may need to manage unrestricted regrowth as it has sprawl. That would mean a stronger hand in the woods, and drip-torches applied for ecological purposes not just to flash off surface combustibles. All this would require a surer sense of what the reserve is about and how it relates to its sustaining society beyond the provision of open space.

As with the pine plains, there are two visions of the recovered forests, both based on nonanthropocentric values. One looks to the National Wilderness Preservation Act, to the wild in its untrammeled transcendent wonder, and is prepared to let nature take whatever course it chooses. If casino gambling proposed one future for an unprotected Pinelands, the simultaneous fight over Alaskan wilderness offered an alternative for a protected Pinelands. It points to eruptive fires that sooner or later swipe landscapes clean in a recurring Götterdämmerung. The other vision looks to biodiversity, as encoded in the Endangered Species Act, and recognizes that the nonanthropocentric can only thrive in an anthropogenic world if people intervene. It points to routine burning, not only for fuel reduction but for ecological engineering. Under the existing program, ecological benefits are a welcome collateral outcome to prescribed burning. Ideally, that relationship could be reversed, such that fuel reduction would be a side-product of burning aimed at delivering ecological goods and services. The prevailing assumption is that land use should determine fire regimes. Given the long history of landscape-scale burning, however, it might make more sense to assume that the fire history of the Pinelands will, over the long term, determine land use even if fire does not occupy a chair at the Pinelands Commission. The New Jersey Forest Fire Service is clear sighted about what it can and can't do. It simply seeks to boost the odds it faces. But the house odds favor the big fire. The hazards have increased, on both sides of the wildland-urban divide; ignitions have remained constant, with a dark shadow cast by arson; only ceaseless vigilance keeps the scene from exploding. Sooner or later a monster fire will clear the table. The catastrophic fire of the future may not, however, resemble those of the past; not the 2007 Greenwood burn that spared the village of Whiting through a providential wind shift, not the 1995 Warren Grove fire that brushed against Stafford Forge, not the 1963 complex, not the 1930 rampage, not the fires that almost annually in the late 19th century swept back and forth across the Pinelands like sea and land breezes. The catastrophe may more closely follow the outbreaks that blasted through Bastrop County, Texas, and Gatlinburg, Tennessee, their smoke plumes within sight of the Texas Capitol and Dollywood, respectively, both feeding on a similar tangle of wildland and exurb, state parks and private holdings, a unique pinery overgrown by houses, oak, and understory, unhinged by drought and pummeled by a blistering wind. It will result not from a single spark that blows up but from many ignitions that overwhelm initial response and merge.

The Pinelands woods continue to recover—that's the good news. The bad news is that it will likely assume forms unlike those of the past. Gypsy moths have stripped oaks, bark beetles are killing pine, drought is upsetting the water regime, climate change is unhinging the tempo of fire weather, feral greenbrier and mountain laurel are overrunning unburned forests, the woods are choked with hydrocarbons like a toxic dump— sooner or later southern New Jersey may know the fire equivalent of a Hurricane Sandy. A Category 3 or 4 wildfire would radically restructure not only its physical geography but its political landscape. A revolution would only require one such event.

———————

The narrative of Pinelands exploitation and abandonment has its doppelgänger in a narrative of attention and forgetting. Over the past century its fires have commanded significance, even national concern, only to sink in the bogs and sugar sands.

Gifford Pinchot and Henry Graves made the pine barrens into an exemplar of bad burning, a miniaturization of what was wrong with land use across the country, and then reasoned that fire was their best bet for galvanizing public opinion in favor of protecting the forested estate of the public domain. They immediately made the issue a national one. New Jersey became a test case for installing modern forestry founded on fire control. Then attention wandered north to the Lake States and west to the nation's vast forest reserves. After his own survey of the pine barrens, Harold Lutz went to Alaska, where he published a comparable survey that influenced fire protection as that vast territory headed into statehood.

In the 1930s, building on the protective burning begun by the NJFFS in 1928, Silas Little and E. M. Moore of the Lebanon Experimental Forest undertook a series of meticulous field experiments unprecedented in American experience. Little began his trials a year after the U.S. Forest Service adopted the 10 a.m. policy for universal suppression, and argued that prescribed fire was useful—this some seven years before the USFS allowed the practice on the Florida National Forest. He published his results nationally—in the *Journal of Forestry*, no less—in the 1940s, a northeastern version of what Harold Weaver was doing, with far less methodological rigor, on Indian reservations in Arizona and Oregon. He spoke to the third Tall Timbers Fire Ecology Conference, giving the Pinelands and prescribed fire in the Northeast a voice in the national discourse that would result in a revolution of fire policy. In 1974 he contributed one of six regional chapters in the first fire ecology text published in the United States, speaking on the same podium as C. E. Ahlgren, E. V. Komarek, Harold Weaver, Harold Biswell, and Robert Humphrey.[14]

Silas Little was in fact a contemporary of Ed Komarek; the Lebanon Experimental Forest was a counterpart to the Cooperative Quail Study and its successor, the Tall Timbers Research Station; both men and institutions were founded out of a concern with fire applied on the land, and both supplied an empirical and conceptual foundation for its use. Yet Tall Timbers became an international clearing house for fire science and a megaphone for policy reform, while the Lebanon Forest sank into obscurity. After the Pinelands Protection Act passed, Little retired and the forest was dissolved. The obscurity of the pine barrens within the American fire community has its counterpart in the lost legacy of Si Little.[15]

Now fire and the Pinelands are back on the table. With funding from the National Fire Plan, the experimental forest resuscitated its fire program in 2002 with research by the U.S. Forest Service and Rutgers University. There are discussions about reforming the New Jersey Forest Fire Service, and a new prescribed fire law was enacted in 2018. Over it all, like a pall from the past, hangs the horror of another breakout fire, this time devouring an entire community. It's a good occasion to reconsider the place of fire in the Pinelands, and the Pinelands in the national fire narrative. This time, if the right people chose, it could become a regional hearth for fire science and a national firepower. The nation's primary fire triangle—Florida, California, and the Northern Rockies—could become a more balanced rectangle.

All the necessary pieces are present. The Pinelands have an indisputably fire-prone ecosystem, one for which fire can only be excluded by forcibly stripping off the biota. They have an unbroken fire culture, passed through generations, rooted in the land and the pineys who live there. Its fire agencies are adept at both suppressing and prescribe-burning. The NJFFS's skills at crafting gear argue for a northeastern companion to the Lake States' Roscommon Equipment Center. In the pitch pine they have an emblematic species, a Northeast equivalent to the longleaf, ponderosa, or sequoia, and in the pine plains, a landscape as iconic as the Everglades or Southern California chaparral. A research capability, complete with its own field station and legacy, has revived; the Pinelands could become the field station for prescribed fire throughout the coastal barrens and its backcountry. The region has breadth and variety—big enough to tolerate considerable experiments, diverse enough to be interesting, close enough to major population centers to be visible and politically compelling. The Pinelands are among a handful of places that have the right constituents, even if they rest together like marbles in a bag rather than valencing like electrons into a new molecule.

Critically, it has the Pinelands Commission. What most prospective hearths lack is an institutional infrastructure. They can imagine, for purposes of fire management, cutting the fences that divide a common landscape into separate political jurisdictions. They can't herd the many constituencies into a common corral, have no mechanism to encourage or compel discussion, can appeal to no mechanism to bring discourse to a collective decision. The Pinelands Commission does. With some clever tweaking it could provide what most other regional campaigns lack: it could create an institutional landscape to overlay its geographic one. In most places efforts stumble and stagger from project to project, stepping in place rather than striding down the road. A Pinelands fire consortium could begin where most regions strive to end. It could do for the Northeast what Tall Timbers has done for the Southeast.

The region deserves a focus. Without federal lands, the national contribution gets funneled through such means as sporadic research, grants in aid to states and volunteer fire departments, and excess equipment transfers. The Pinelands could reach well beyond the Jersey shore. The pine barrens extend north and south; they support fires on Staten Island and Long Island, along Cape Cod, and even into southern Maine. Since 1949 the northeastern states have formed a compact for mutual assistance in fire suppression; they need a comparable one for prescribed burning. Some controlled burning occurs sporadically along the coast and, thanks to the Nature Conservancy, on Martha's Vineyard and outside Albany for the Karner's blue butterfly. The big states, New York and Pennsylvania, have lagged (Pennsylvania only conducted its first legal prescribed fire in 2010). The region lacks a strong-force nucleus, one grounded in field science, to roughly hold its electron swarm of institutions. Moreover, the Pinelands offer a complexity of intermixed wild and urban sites that make western equivalents seem cartoonish. Yet this is the characteristic landscape of the eastern United States, and if climate change models are anywhere near accurate, damaging fires will become more prevalent. The Pinelands could host the scene for alternative experiments in Firewise and for institutional arrangements that do not depend on federal agencies for the gravitational attraction needed to hold them together.

The national fire scene, too, could use another anchor point. The National Cohesive Strategy divides the country into three regions. The Southeast and West have fire clusters like pyric Silicon Valleys of research, equipment development, and high-volume fire activity that create synergy and suffuse a characteristic style throughout the larger region. The Northeast does not. From time to time an effort to create a northeastern fire presence flares, then fades. Yet the idea is a sound one. Like urban parks that seek to bring parklands and their agencies to the public, a northeastern fire cluster could carry fire issues close to where a more major fraction of the American population lives.

════════════

A northeastern center would have its own peculiar flavor, bolted to enduring themes in the region. In the West, fire bonds to wilderness and public land; in the Southeast, to working landscapes and private practitioners as well. In the Northeast much of the public land belongs to the states, and many working landscapes have converted to sprawl or recreational usage. While the National Fire Plan, and its successor, the National Cohesive Strategy, attempt to unify all three regions through an emphasis on fuel treatments, the effort has faltered both conceptually and practically because a dissonance exists between a formal emphasis on fuel treatments and what society wants from those lands in terms of ecological goods and services and such cultural values as wilderness. Here and there restoration forestry has combined a fuels-informed thinning with burning to enhance landscape health. In the Northeast, however, fuels may be where fire science, social expectations, and ecological needs converge.

Certainly from the time of European contact, the Pinelands have repeatedly been assessed as and reduced to fuel, whether burned on open plains or in ovens, furnaces, and steam engines. That history selected for the pitch pine not only as a survivor of serial conflagration but as itself an energy-rich combustible for home and factory. Even the tradition of prescribed burning was framed in terms of hazard reduction: it burned under controlled conditions what would otherwise combust as wildfire. Such attention only shifted when alternative combustibles, in the form of fossil fuels, could power the human economy. For the first time in four centuries the Pinelands are not primarily a woodlot: they can be valued for social amenities and ecological goods and services, and burned for biological benefits, not just hazard reduction.

Such an outcome seems unlikely any time soon. What the Pinelands could do is to show how, through fuels management, it might be possible to achieve those other collateral values of the land. Getting the fuel array right, by reducing the danger of devouring fires, would grant space for other purposes. It could also allow for the systematic study of the two realms of fire that define the pyrogeography of the Earth today—the open burning of living biomass and the internal burning of fossil biomass. Rarely have they been linked conceptually, much less had their interaction formally scrutinized, yet that dynamic is what drives the national fire scene. In most places fuel makes a crummy metric by which to describe ecosystems; yet in the Pinelands it has a historical logic. Since European contact its forest has always been defined as fuel. A fuel-centric research could show, for example, how to expand the opportunities for burning, now limited to 15 to 20 days between January and mid-March, by shifting the focus from seasonal calendars to days of fuel availability. The pitch pine, and its associates, could supply the needles and windfall combustibles to permit fire to perform its ecological duties. They could fuel a biota.[16]

If the fire ecology isn't exactly right, that's also because the forests are still rebounding and sorting themselves out, and because they exist within a matrix no longer simply inscribed by wet bog and dry woods but by open burning wildland and internal-combusting city. The former threatens to blast landscapes with a flaming front; the latter, to pave over the countryside and slow-cook the planet. The grand task before fire management is to bring them into alignment so that they enhance, not eliminate, the landscapes on which they flourish. If their fuels run amok, so will their fires, and whether they burn fast or slow, the consuming flames will devour them all.

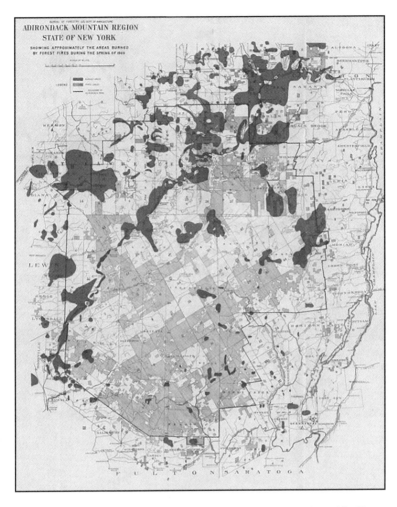

FIGURE 1 Map of the 1903 Adirondack forest fire. From H. M. Suter, The Forest Fires in the Adirondacks in 1903 (dark patches).

FIGURE 2 Adirondacks firefighters, ca. 1898. Photo from U.S. Forest Service Historic Photo Collection.

FIGURE 3 New Hampshire fire, Pemigewasset Valley. Photo from U.S. Forest
Service Historic Photo Collection.

FIGURE 4 Logged and burned—"scalped," in President Theodore Roosevelt's words, Manistee National Forest, 1901, four years before the U.S. Forest Service assumed control over the reserves. Photo from U.S. Forest Service Historic Photo Collection.

FIGURE 5 Aerial view of 1947 Maine fires. Photo from U.S. Forest Service Historic Photo Collection.

FIGURE 6 Ground view of 1947 Maine fires. Photo from U.S. Forest Service Historic Photo Collection.

FIGURE 7 The pine plains, the heart of the Pine Barrens. Photo by Stephen Pyne.

FIGURE 8 Prescribed fire in the Pinelands, March 2013. Photo courtesy Bob Williams.

FIGURE 9 Allegheny Plateau, 1926. The case for fire protection was easily envisioned. Photo from U.S. Forest Service Historic Photo Collection.

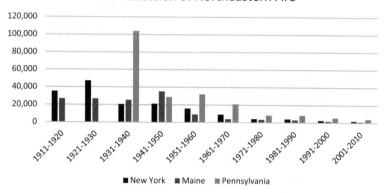

FIGURE 10 New England fire history by decade, as represented by New York, Pennsylvania, and Maine. Data from Lloyd Irland, "Fire History of New York State, 1910–2010"; Pennsylvania Department of Conservation and Natural Resources; Lloyd Irland, "Maine Forest Fire History and Analysis 1903–2010."

FIRE ON THE MOUNTAIN

I N NATURE, in culture, in the ways they come together—the Adirondacks are an oddity.

Geologically, they are not part of the Appalachians—the only eastern mountains that stand apart. They are a fragment of the Precambrian Shield that welled upward about five million years ago into a dome 160 miles wide and a mile high, rising out of the shield like the Black Hills from the Great Plains. For 2.5 million years they eroded under the influence of seasonal weather, the rhythms of summer and winter. For the last 2.5 million they have been sculpted by the planetary rhythms of glacial and interglacial that left them gouged, rounded, abraded, and polished. Today they form a bubble island, set within a rough triangle of waters; a chain of glacial lakes to the east, the Mohawk River to the south, the Great Lakes along the hypotenuse. But their peculiar geology is only a point of departure.

Their most striking oddity is that they comprise the largest American park outside Alaska, three times the size of Yellowstone, but a park quilted from public and private land, with purposes ranging from preservation to private recreation to outright commercialization. The Adirondacks are a new mountain made of old rock; Adirondacks Park is a new idea composed of old notions. It shows an alternative history to what was, at nearly the same time, emerging out West. As New York City stood for the commercial history of the United States, so New York State stood for its environmental history. It proposed an alternative pathway—for parks, for forests, for fire—to federal oversight of an expansive public domain.

Today the Adirondacks have little fire. What they had in their geologic past is unknown, and what they had at the time of European contact is unclear. Lying within the temperate Northeast, without sharp rhythms of wetting and drying, with sparse lightning ignitions and little permanent human habitation, fire was likely infrequent, if sporadically influential. But with European settlement the Adirondacks acquired fire aplenty: fire passed over the mountains historically as a traveler might pass over the rise and fall of the mountains themselves. It was the old tag team of fire and axe that kindled those burns and then inspired the movement that set the mountains aside as protected land, and by its example such sites as Mount Katahdin in Maine and even the ramparts of the Rockies.

The Adirondacks were odd, too, because they were equally far and near—close enough to power to feel its influence, remote enough to draw as a natural spectacle. They were not remote, however, from the national narrative of fire.

———

Settlement came late. The lakes to the east were an ancient thoroughfare between the northern Algonquians and the southerly Iroquois Confederacy; it kept that status during the century-long fight over trade and empire between French Canada and Dutch (later, British) New England. The Mohawk River was a similar path westward, later connected directly to the Great Lakes by the Erie Canal. The westward migration flowed around the mountains like a stream around a massive boulder. Once the beaver were trapped out, there was little settlement. Logging and farming gnawed at the margins.

With clearing came burning. There were far more accessible forests, however, and the Adirondacks' rivers were not easily converted to flumes for floating logs. The first wave of forest felling over New England mostly passed the Adirondacks. But the second wave, with steam and rail to crack open the interior, did not. Axe chewed into the white pine, and fire burned through the slash and into whatever surrounded it. The Adirondacks began to burn, occasionally prodigiously. The flames licked at their margins, then, as the settlement moved inward, advanced on the mountains.

They even found their way into American literature, as James Feni-more Cooper brought *The Pioneers*, published in 1823, to a climax with a fire on the slopes of a burning Mount Vision that gathers together a young couple, Edward and Elizabeth, who represent the future of a set-tlement beyond raw stumps, and two aged men of the mountains, Natty Bumppo and Chingachgook, who come from a more prelapsarian time before the axe bit into the woods. The fire results from slovenliness, the hard carelessness by which the frontier advanced: slopes littered with "tops" from felled trees, a miner indifferently tossing lighted pine knots into the brush. It is not a world in which the old Mohegan and Pathfinder can continue. But the panorama from Mount Vision is suddenly obscured by "immense volumes of smoke" that rolled over their heads and roared with the winds, and by trapping Edward and Elizabeth the burn threatens the prospects for a wiser, more genteel, less ruinous way with the land.

When he learns of the danger, Natty sends a messenger "to the village" to give the alarm. "The men are used to fighting fire." Then he plunges into the smoke, emerges "without his deerskin cap, his hair burnt to his head, his shirt, of country check, black and filled with holes, and his red features of a deeper color than ever, by the heat he had encountered." He rescues the young couple and his old companion, Chingachgook, hauling him on his back. Then, suddenly, lightning cracked, an "awful stillness . . . pervaded the air," and rains poured to quench the fire. "The next day the woods for many miles were black and smoking, but were stripped of every vestige of brush and dead wood; but the pines and hemlocks still reared their head proudly among the hills." In the end, a kind of frontier justice prevails. The young couple are united; the miner who started the fire is found nearly suffocated "in his hole," and so burned he will die; Chingachgook goes to his reward before he must see the final ruin of his heritage; Natty heads West; the land regrows into something between the wild and wrecked.[1]

In *The Pioneers* Cooper gave American literature its first sustained account of a forest fire, its first entrapment, its first rescue (using a buck-skin coat as a shelter), its first fire fatality. Cooper was famously lampooned by Mark Twain for his "literary offenses" and fanciful woodsmanship, and even at this point—his second novel—Cooper felt the need to add a footnote to address complaints about its implausibility. Cooper replied that he had "once witnessed" such a fire in another part of New York, and that the likelihood of such fires was far from remote. He enjoined readers "to remember" the effects of drought and the "abundance" of forest slash. "The fires in the American forests frequently rage to such an extent as to produce a sensible effect on the atmosphere at a distance of fifty miles. Houses, barns, and fences are quite commonly swept away in their course." In the last of the *Leatherstocking Tales*, Cooper dispatches the Old Trapper to the prairie, where he casually burns through the dead grass— another first in American fiction.[2]

The settlers who did most of the work did not reap what they hoped for. The wreckage harmed all (the *Pioneers* includes a vivid account of slaughtering passenger pigeons, among other environmental insults). Along the Adirondacks, the small farmer was typically a big failure. "Probably in no other part of the United States, certainly in no other part of the State of New York, can a more unfortunate agricultural population be found." The legacy of landscape-converting fires was too often collective degradation.[3]

In his 1880 report Charles Sargent observed that the lumber industry of the upper Hudson could continue indefinitely "if the forest could be protected from devastating fires." But loggers left slash, and in "over at least one-half of the area lumbered fire follows the axe, burning deep into the woody soil and inducing an entire change of tree covering." The fires were "largely set by reckless sportsmen and hunters, with whom this region peculiarly abounds in summer." They were careless with their campfires, and "sometimes they even delight to set fire to the dry brushwood of lumbered land in lawless sport." The fires spread beyond the "lumbered" lands. "The laws of New York in respect to the setting of forest fires are totally inadequate to protect the forests."[4]

Meanwhile, another settler began to arrive, this one seasonal and interested less in commerce than in recreation. The Hudson River was a corridor for artists as well as trappers and soldiers; and following the artists were those who found in the mountains the ideal trappings of a summer home, or hunting and fishing camps, or sanitariums. For the wealthy, the Adirondacks became the inland counterpart to the Connecticut shoreline. Rail began to haul tourists as well as logs. The Adirondacks promised the scenic splendor of Far West nature within a day's journey; they were the largest block of uncut forest in the Northeast. What the wilderness West was to the country, the Adirondacks were to New England.

Opening the mountains opened them to all. The axe met second-home cottage, resort hotel, and millionaire mansion. Thoughtful partisans of the mountains appreciated how the combination of logging and touristing could hollow out the grandeur of the mountains, like fire and termites working in tandem. Worse, the fires were morphing from nuisance to threat. Together, axe and fire were the industrial-strength scissors that threatened to cut up the mountain forests, wreck the watersheds on which the state's arterial waterways depended, and trash the scenic viewsheds on which the second of the region's economies, tourism, relied.

The saga of how an Adirondacks park was created proceeded in syncopation with those in the West. It had the same arguments and alarms: fire and axe threatened the land with ruin. It appealed to the same mechanisms: state-sponsored conservation in which government, in this case the State of New York, had to intervene to halt the depredations of migratory capitalism and monopoly. It even enjoyed the same timing. In 1884 a California court shut down hydraulic mining in the Sierra Nevada. In 1885, when the Adirondacks commission issued its report leading to the Adirondacks Preserve, Ontario established its first corps of "fire rangers" to tackle abusive burning on crown lands. In 1886, as the Adirondacks mobilized to operationalize fire protection by way of county fire wardens, the U.S. Cavalry rode into Yellowstone National Park to bring the kind of systematic protection the unfunded civilian administration had failed to achieve. M Troop, U.S. First Cavalry, extinguished 60 fires that first summer.

─────────

For years advocates like Verplanck Colvin and Franklin Hough had warned about the growing menace. Public concern first crested in 1872 with the creation of a State Parks Commission to inquire about the feasibility of an Adirondack reserve—this was the same year Congress established Yellowstone National Park. The commission recommended, at a minimum, regulation over wasteful logging; two years later, Colvin urged the state to fashion a forest reservation out of still-extant state lands. In 1876, in a separate action, the legislature updated fire laws (last done in 1788) to criminalize escape fires that spread to the property of another and authorized select officials to impress people to fight fires. The logging industry of course objected to any constraints. But a coalition of commerce and intellectuals, the elites of industry, government, and culture, congealed to halt the havoc, as a similar alliance had in California to end hydraulic mining and protect sequoia groves in the Sierra Nevada.[5]

The political response was to authorize a forest commission, chaired by Charles S. Sargent, then a professor at Harvard and director of the Arnold Arboretum, previously tasked with a survey of the nation's forests for the 1880 census, to report on the status of New York's forests, with special attention to the Catskills and Adirondacks. The commission delivered its report to the state comptroller in 1884, who forwarded it to the legislature on January 23, 1885. It recommended state management of New York's remaining public forests. Its concerns were not merely for timber or recreation ("this great lucrative business") but for the effect of forest clearing on climate and watersheds. The Sargent Report remained for years the standard political reference on matters Adirondack.[6]

The commission's report recognized the complexity of the scene and subtleties that a political solution would require. But it softened no words regarding fires, which were "slowly and surely destroying the Adirondack forests. Unless they can be stopped or greatly reduced in number and violence, nothing can prevent the entire extermination of these forests." The causes were as varied as the people tramping through and around the woods—campfires, smudge pots, charcoaling operations, locomotives, farmers, and fallow. Each fire created the conditions for another, until the multiple reburns left "denuded" patches. The perimeters of the remnant forests were becoming a "desert belt" whose relentless push inward was gradually compressing and squeezing the life out of the Adirondacks. "What fire has done in the past it will do in the future, and so long as these fires are allowed to rage, nothing can prevent the extermination, at no very distant day, of the whole Adirondack forest. The preservation of this forest is reduced to a question of the possibility of preventing forest fires. If they can be stopped, this forest can be preserved; if they are allowed to go on, nothing can prevent its early and total ruin." These were essentially the same terms applied, in often identical language, to the national estate a decade later by the National Academy of Sciences Committee on Forests.[7]

Within the year legislation based on the Sargent Report, with further input from foresters (including possibly Bernhard Fernow), established a forest preserve with language that would underwrite the future of both the Adirondacks and the Catskills. "The lands now or hereafter constituting the Forest Preserve shall be forever kept as wild forest lands. They shall not be sold, nor shall they be leased or taken by any person or corporation, public or private." Chapter 283 of the Laws of 1885 was signed by Governor David B. Hill on May 15, 1885—the same year that the American Forestry Association held its first national convention, six years before the National Forest Reserve Act, and 20 years before the creation of the U.S. Forest Service. A commemorative history by New York identified it as "the fundamental conservation law of the State and the first comprehensive environmental law in the nation." A strong claim, but one that demonstrates the significance granted it by the region. Thanks to Sargent the Adirondacks Preserve was clearly the model when the time came to plan for national reserves. Even the language of their informing questions was nearly identical.[8]

The law established a forest commission, which laid down regulations for preventing trespass by axe and fire. The principle mechanism was a system of fire wardens, one to each of the counties within the preserve. The wardens were to divide each township into fire districts; they were charged with rallying townsfolk to fight fires; they could enforce laws, for example, those requiring locomotives to have spark-arresting screens. The wardens were not foresters: they were charged with fire protection. But fire duties so absorbed the early years that fire wardens were the nucleus of a forest ranger force.[9]

There was plenty of fire to keep them busy.

In its first report the commission elaborated on the menagerie of causes. Foremost was agriculture: farmers burned to clear new land, to recycle fallow ("fallor"), to improve the natural meadows left after beaver ponds filled in. These were typically set in April to get a hot fire, and not infrequently bolted into the nearby woods. Berry pickers left cooking fires and burned the shrubs every few years. Gum hunters searching for spruce gum ("quite an industry") left campfires, often in remote settings; some of the fires "have left marks never to be obliterated." Bee hunters, too, left cooking fires, but added fires set to smoke bees out of trees, and unless "water is handy," did not bother to extinguish them, trading "a few pounds of honey" for "many acres of valuable timber." Hunters set fires to ease passage through brush and brambles (to say nothing of logging slash), and to encourage the "buckhorn" that attracted deer. Iron manufacturers harvested "large sections of timber" annually, and burned off the debris; their fires did not always stay on the cutover patches. "Evil-disposed persons" started fires at the foot of mountains, "just for the fun of the thing, to see a big blaze," an admittedly "grand and awful sight"; burned sites "out of revenge for real or fancied wrongs"; and fired slash to destroy evidence of illicit timber harvesting. Add to these sheer carelessness. Wherever people congregated, they lit fires for "cooking, warmth and light," often in "exposed places" where they could smolder and gnaw down into deep duff and flare up days or weeks later. Wherever they stopped, they kindled smudge fires to "drive off gnats and flies." Others set "big, roaring, out-door log" fires of the kind impossible to experience in a city. Even conscientious visitors cleaned camps by burning refuse and brush, a practice that "has been the cause of denuding the shores of some of the most picturesque lakes and ponds in the Adirondacks." As railways pushed inland, locomotive fires trailed behind as surely as the smoke from the engines. The report for 1896 added smokers ("more numerous and destructive than would appear at first thought") and natural causes ("extremely rare, and hardly worth mentioning"). And always there was the odd-bins bracket, "Unknown cause," although it could be breathtakingly large (36 out of 94 fires in 1896). What added accelerant to the ignitions was the raw rash of logging slash. The Adirondacks were not naturally disposed to burn, or burned in small patches. Industrial logging littered the landscape with debris of exactly the right sort and arrangement to burn. The "combined results from these various causes" was to turn routes of travel into fire-denuded corridors and to dapple the mountains with growing cancers of slash-and-burned fields of fire. What mattered especially was stopping the cycle of reburns.[10]

Against this threat the commission established a fire warden system. It was a simple institution, built on early traditions in which an official could call on ("warn") the able-bodied to fight fire (for pay). Locals could handle most fires, as Col. William Fox, assistant to the secretary of the commission (later himself superintendent of forests), explained. "The provision of the law requiring a statement as to the means used in fighting or extinguishing the fire was generally complied with by the firewardens in their reports," he reported in 1886.

From the information thus furnished it appears that the common method used in stopping a small fire was by whipping it out with brush, after which fresh dirt was thrown on the smoking leaves and embers; or, if there was water near by, it was carried in buckets and poured over the ground. Where the fire had gained a good headway and spread beyond control, a line of defence was chosen along some road or stream, from which back fires were started. In some places where there was a slow ground fire, and the soil would permit it, furrows were plowed, and a space was swept bare of leaves and combustible material, thereby making a line at which the creeping flames stopped for lack of fuel. In most cases where the fire covers a large area, the men work in the early and late hours of the day, or in the night; for then the flames die down, and can be fought easier than in the daytime or noon hours, during which they burn with uncontrollable fury.

In the case of a "top" fire, driven by a strong wind, little or nothing can be done to stop it, aside from extinguishing the small fires that start on all sides, lighted by falling sparks or brands. Sometimes, a sufficiently large posse of men having been warned out, a top fire has been encircled by a cordon of fire fighters, back fires started, and the conflagration held in check until rain came to their relief. With few exceptions, our larger forest fires in the Adirondack and Catskill regions burn until rain comes. Fortunately, in the spring and fall, the times when all our woodland fires occur, the rains are most frequent, especially on the mountain plateaus where the forests are situated; and a fire seldom lasts four days without being extinguished by some opportune downpour. The frequency with which showers follow forest fires has led to a prevalent belief in the certainty of this phenomenal succession as an ordinary exhibition of cause and effect. Rain is the best firewarden we have, and were it not for this agency there would be no forests to-day on the Adirondack and Catskill uplands.[11]

The genius of the warden system was that it drew from tradition and town fire departments and had a pool of labor that was not merely impressed but paid (half by the town, half by the state). The deficiency of the system was that it could only respond, not patrol and prevent; it relied on the towns to pay for their share (the town could prove surly and delay indefinitely); it could not install an infrastructure (including protection roads and fuelbreaks); and it was unable to handle any extended attack. It stopped many fires from becoming big. It could not do much with those that did blow up.

The warden system was a major improvement. "Prior to 1885," Fox noted, "it was a common event, an almost annual occurrence throughout the state, to have the atmosphere obscured by the smoke from burning woods—a thick, blue haze through which the sun at noonday appeared like a dull, red disc. This phenomenon is no longer seen in our State except as a local condition." Already the commission's efforts had demonstrated that "forest fires can be materially reduced in frequency and extent." It was possible.

Still, the overall fire scene continued to deteriorate. Those big fires were the big problem—the ones that did the most damage and once started could not be beaten back. The fire warden system could not scale up as large and as quickly as the sources of the fires and their enablers. Then the system collapsed. "The year 1899 will long be remembered by the people of our North Woods as the season" when the fires grew great. An "extraordinary" drought powered an extraordinary outbreak of fires that "in number and area, far exceeded any that had ever happened in all that region."[12]

Accounts of the fires are unusually rich. Commissioner Charles H. Babcock was himself in the Adirondacks at the time, as was the superintendent of forests. Assessing the scene, Babcock hurried to Albany for a "personal consultation" with Governor Theodore Roosevelt and the state comptroller, Col. William J. Morgan. The fires were many and wide-spreading, but because they broke out in August, with the trees "in full leaf," the damage to untouched woods was less than the statistics might suggest. Most of the fires burned on clearings and wastelands that had been burned over "once or twice before." The fear was that the flames would eventually rip into the uncut forest and might continue into autumn long enough to gain new strength as the leaves fell.[13]

The alarm was the inability of the protection system to respond. The commissioner explained, from firsthand knowledge, the problem.

In many of the towns there was an evident reluctance on the part of the officials to warn out the necessary number of men to fight fires owing to the heavy expense and consequent taxes which would be assessed upon the town. In some towns, also, where the authorities were willing to do whatever was required, there proved to be a scarcity of men. In other places men refused to go to a fire, alleging that would not get their pay from the town until after the annual meeting of the board of supervisors, which occurs in December; and that even then they would receive no money, but, instead, an order on the tax collector, which they would have to sell at a discount.[14]

The upshot was a crisis. With "commendable promptness" Governor Roosevelt and Comptroller Morgan agreed to set up an emergency fund to pay those men called out by the wardens, and in remote settings, where there were no local residents, to hire men from the region outside and transport them by railroad. In fact, a special train was arranged that worked for four days, "both night and day." Nine years later, the Roosevelt administration oversaw legislation to create an emergency fire fund for use on the national forests. The quirky financing of wildland fire had its origins in the Great Fire of 1899 and its make-or-break test during the Great Fires of 1910 when the overdrafts for the U.S. Forest Service ran to nearly a million dollars.[15]

The supposedly great fires that closed the century were soon overwhelmed by others far worse. The decade beginning with 1899 marked the crest of the Adirondacks' Era of Great Fires. Superintendent Fox argued for the creation of a corps of rangers to do what the wardens could not. Then came the dual fire years that shattered the old system.

From April 20 to June 8, 1903, fires swarmed over 600,000 acres of northern New York, of which 643 fires and 428,180 acres were in the Adirondacks. Unlike 1899 this was prime fire season, atop a record drought, and the fires did not just reburn old sites but were carried by rail and transients throughout the mountains. Locomotives scattered sparks, railway rights-of-way acted like fuses, and plateaus of slash blew up. Matching the trains were recreationists, particularly fishermen, who littered fires with abandon. A map of the burns shows them tracking railways and streams, the swathes wending to where logging debris gathered. Once started the fires sprayed and sprinted wherever the winds took them. Ash fell on Albany. Darkness spread over New York City. The smoke was felt in Washington, D.C.

There was no break in fighting the fires. Everyone who could be "warned out" was, many of the men more than once. The firefight racked up a total of 77,290 man-days. Towns rallied and rallied again for their own protection. Special fire trains hauled laborers in from as far as New York City. A freight car was outfitted with a special pump to quench fires along the right-of-way. Flatbeds carried city fire engines from unthreatened communities nearby. Between thickening smoke and fatigue, fire control faltered as the weeks passed. "The destruction of the entire region seemed not at all improbable, for in the dense pall of smoke it was impossible to tell where the fires were." In some places the fires could be heard and not seen. Some nights seemed "almost as bright as the days from the glare." The U.S. Division of Forestry under Gifford Pinchot dispatched Herman M. Suter for a detailed report. The cost of the fires amounted to $175,763.95, half of which was the responsibility of the local towns, which could ill afford it. Total loss was estimated at $3,500,000. Incredibly, no lives were lost. Suter concluded that "the blame for the avoidable loss lies rather with the system than with the men."[16]

What couldn't get worse did. In 1908 the spring fires flipped into autumn. It was a season of fire nationwide. The fall burns had rolled across the northern tier of the country from Washington State to Maine. Everything that made for big fires in the Adirondacks converged with almost preternatural cunning. Everything scrambled to a peak—drought, logging, railroad construction, the overextension of the warden system. When the rain and snow finally quenched the flames, a total of 368,000 acres had burned, much of it wrecked. This was not a forest that relied on regular burns for rejuvenation; the fires, stoked by logging slash 12 to 15 feet high, were far from regular or natural.

The worst single fire broke out near Long Lake West. Its origin makes a perfect symbol for the season. A spark from a Mohawk & Malone locomotive set a fire along the track. Chief Fire Warden Emmons quickly had 150 men on the scene. In what must stand as an apologue of a vicious cycle, the train bringing them to the scene was itself setting new fires along the route. The town of Long Lake West was soon cinders. By September 11 a line of fire stretched 12 miles from Horseshoe to Nehasane. Multiply that experience for Owl's Head and Belmont and innumerable small communities for 368,000 savagely burned acres and you place the Adirondacks into the continental sweep of 1908's fires.

This time systematic reforms followed. Locomotives were compelled to burn oil rather than wood or coal during fire season. Loggers were required to limb their slashings. A permanent infrastructure arose built around fire towers and a staff of patrolmen. The governor was authorized to close forest lands to entry during high fire danger. Over the years the frenetic cords that had knotted together to kindle the great fires began to unwind.

The national fire narrative focuses, rightly, on the great fires of the West, particularly those of 1902 and 1910. For the Northeast the critical seasons were 1903 and 1908. A case can be made, though, that the Big Blowup of 1910, while formative for the Forest Service, was a political echo of 1903 and 1908. The Big Blowup in the Northern Rockies turned the sky over Boston a coppery yellow and made the front page of the Spokane *Daily Chronicle*. The Big Blowup in the Adirondacks smothered New York and Boston, and produced headlines in major media that regional politicians could not ignore. The smoke was felt, literally, in the halls of Congress. It lingered in the memory of men who would be in a position to decide fire policy nationwide.

======

The fever broke. More equipment, more towers, more rangers, more railroad inspectors, more laws to regulate what could burn and when, a fundamental reorganization of the warden system, the whole apparatus by which New England overall had leached fire out of the mountains came into play. The Adirondacks are a miniature of the New England fire narrative: fire rose and fell for the same reasons it ramped up, and it declined likewise. Through bureaucratic reorganizations, through drought and blowdown and insect outbreaks, the number of fires and the amount of area burned dropped. Some years defied the average, and there were troubling outbreaks like those in the mid-1960s that spanned several years, but each crest was lower than the previous until fire seemed to flatline. It flared, as an outlier, in pitch pine-scrub oak expanses like those at Sam's

Point Preserve and in eastern Long Island, not in the mountains. So much fire had been removed that critical habitats suffered from too little. The fire scene of the early 20th century had been turned upside down. In 1970, amid a fire revolution in the United States that sought to restore landscape burning, New York banned open burning in forests.[17]

The Adirondacks were the backyard for many of the politicians and intellectuals who shaped national fire policy. They saw fire as a social problem—a problem of human behavior—and they systematically reduced it to the point that it threatened to become of archaeological interest. It was only natural for them to assume that the Adirondacks—the last quasi-wilderness in the Northeast—was a model for the vaster wildlands of the American West. They could reduce the fire hazard by the same methods. The Adirondacks were proof of concept. It had worked in New York. It could work anywhere.

For Charles Sargent the forest commission was a dress rehearsal for the National Academy of Sciences Committee on Forests two years later. For Teddy Roosevelt, forest conservation in the Adirondacks foreshadowed his transformative labors as president. "All that later I strove for in the nation in connection with conservation was foreshadowed by what I strove to obtain for New York State when I was governor." Franklin Roosevelt, too, learned forest conservation as governor. The country could be salvaged from axe and fire as the Adirondacks had.[18]

Unthinking people had brought fire to the Adirondacks. Thoughtful people had removed it. More thoughtful people will have to find the means and ends to put it selectively back.

―――――――――――

The future of the Adirondacks is uncertain, as the future is uncertain everywhere. The park seems as politically secure as any protected site in the country. The national parks exist by act of Congress; the Adirondacks Park, by constitutional provision. How its ecology adapts to the coming and going of species, climate change, the complete extirpation of fire is yet to come.

As a public park it more resembles a biosphere reserve than a wilderness. Its mix of commerce and preservation, recreation and protection makes it look more like the Black Hills or Lake Tahoe than Yosemite or the Northern Cascades. Its reconciliation of public and private interests under a collective planning regime places it squarely in New England traditions and far from the polarized politics of the American West. It survives as the largest park in America and the biggest block of uncut forest in the Northeast. Its iconic imagery depicts high, rounded mountains, mottled with forest, a placid lake at their foot, while in the foreground instead of a mossy boulder, a rotting stump, or a blasted snag sit two Adirondack chairs.

That might also stand for its fire history. The mountains tell a different kind of fire history, a long somnolent era broken by a feral wave of burning that washed over like a Biblical plague and then passed away. That savage moment became a vital moment in the national narrative. Without the threat of fire, which magnified the axe immeasurably, there would likely have been no park. State-sponsored conservation might never have emerged on anything like the scale it did.

Instead the fire program moved West, and like many émigrés it thrived in its new setting far better than in its place of origin. Aggressive fire control was not intended as a permanent occupying force: it was something that would solve the crisis of a historic moment, then fade into background noise, of a piece with road maintenance, nature walks, and game licensing. That proved true for the Northeast. Even today fire restoration is patchy: targeted to species, habitats, places.

But that experience, once exported, misread nature and history in the West where fire was integral and informing, an ecological obligate and broad-spectrum catalyst, whose aggressive removal could unwind landscapes. Remove fire from the Adirondacks and you dampen the conditions that made reburns possible since old burns became sites of future fire infection. Remove fire from much of the West and you amplify the conditions that make conflagrations possible. Outside of grasslands, old burns help to check future blowups.

East and West are different fire worlds. But for a while, in the late 19th and early 20th centuries, they looked the same to many intelligent observers and seemed to beg for the same solutions. Bad, abusive burning had overpowered good burning. Now each has to find some mechanism to restore those lost good fires. The East will do it with targeted prescribed burns in landscapes of high biological value. The West will do it with managed wildfire across complex landscapes. This is as it should be: the essence of fire management is to reconcile fire with land. Different lands, different fires. You don't have to perch on a granite promontory overlooking Yosemite Valley to see nature. You can also view it along a mowed lakeshore from an Adirondack chair.

ALBANY PINE BUSH

MOST NORTHEASTERN FIRESCAPES exist within cultural landscapes. The Albany Pine Bush Preserve sits within an urban one. Adjacent to its fragmented 3,300 acres are tract homes, condos, strip malls, an auto dealership selling Jaguars and Volvos, Walgreen's and CVS, a railway, Interstates 90 and 87, three nursing homes, medical offices, power lines and phone lines, a mattress store, a trailer park, a storage rental facility, Burger King and nail salons, gas stations, Subway and UPS, an Office Max and McDonald's, insurance brokers, engineering consultancies, gyms and massage parlors, a city dump (the highest point in Albany), a VFW hall, and an Italian American Community Center—an herbarium and birding list of contemporary American economic diversity. The Pine Bush Commission consists of the heads of two state agencies, representatives of the City of Albany, towns of Colonie and Guilderland, Albany County, the Nature Conservancy, and four members appointed by the governor. The commission is charged with maintaining or restoring the pine bushlands, a fire-dependent ecosystem, which is the habitat for a constellation of rare species, including the federally listed Karner blue butterfly, which means the pine bush must burn.[1]

As Neil Gifford, conservation director, puts it, if you can burn here, you can burn anywhere.[2]

The Albany pine bush is a sandy plain, the residue of glaciers that spilled into Glacial Lake Albany. Historically it extended between Albany and Schenectady, stocked mostly with pitch pine and scrub oak amid a patchy mosaic of shrubs, grasses, and forbs. Ecologically, it's an inland equivalent of coastal barrier islands. Over the Holocene its fire regime has changed, but something like the contemporary scene seems to have persisted, as recorded in charcoal and pollen, for 4,000 to 6,000 years. With little dry lightning, the ignition source here, as throughout the region's barrens, must have been the indigenes. They had no reason not to burn.[3]

Early Europeans witnessed episodes. In the 1640s Adriaen van der Donck wrote a long account:

> The Indians have a yearly custom (which some Christians have also adopted) of burning the woods, plains and meadows in the fall of the year, when the leaves have fallen, and when the grass and vegetable substances are dry. Those places which are then passed over are fired in the spring in April. This practice is named by us and the Indians, "bush-burning" which is done for several reasons. First, to render hunting easier, as the bush and vegetable growth renders the walking difficult for the hunter, and the crackling of the dry substances betrays him and frightens away the game.
>
> Secondly, to thin out and clear the wood of all dead substances and grass, which grow better the ensuing spring. Thirdly, to circumscribe and enclose the game within the lines of fire, when it is more easily taken, and also because the game is more easily tracked over the burned parts of the woods.
>
> I have seen many instances of wood-burning in the Colony of Rensselaerwyck where there is much pine wood. Those fires appear grand at night from the passing vessels in the river, when the woods are burning on both sides of the same. Then we can see a great distance by the light of the blazing trees, the flames being driven by the wind, and fed by the tops of the trees. But the dead and dying trees remain burning in their standing positions, and appear sublime and beautiful when seen at a distance.

In 1796 Timothy Dwight observed that fires were particularly prone in pitch pine and oak scrub because the sandy soil meant they were typically dry. He speculated that New England's many barrens had been burned for a thousand years.[4]

But those fires didn't vanish with the passing of aboriginal Americans. The colonists often emulated their predecessors, or forged hybrid practices, modifying burning to interact with livestock, introduced cultivars, and land tenure systems. To the old practices such as fire hunting, swidden, and berry production, they added new ones for large-scale landclearing and pastoral burning, and, prodigiously, with locomotives. Farms and fields disciplined burning within a cycle of cultivation; outside those fence lines, it burned without the interactions that had previously characterized it. There is some evidence that clear-cutting, plowing, and slash fires may have shifted the biotic composition. With industrialization the pine bush found new lines of fire along the rails, and ignition ran amok. The world's first passenger steam locomotive made its inaugural run between Albany and Schenectady on August 9, 1831. Its engine burned Lackawanna coal, which it then belched in palls of ash and embers across the countryside. The Albany pine bush found yet another way to burn.[5]

Then the pioneering fires of settlement and steam ended, and after old-field pine had been cut and its slash burned, the reaction known as conservation created formal programs to tame fire within the landscape, and ultimately to cull it out altogether. By 1900 it became policy to abolish fires from the pine bush. Towns established fire codes and enrolled wardens to fight wildfire. States mandated forestry bureaus to assist. Industrialization took people out of rural lands, or if they remained, moved combustion from human hands into machines. Eventually, New York State forbade burning in forests altogether.

For the Albany Pine Bush Preserve (APB) the New York State Thruway and an infrastructure of water, sewers, and electrical power pressed home in the postwar era that suburbanization was imminent, and added a sense of urgency to install urban-style fire protection. When Crossgates Mall opened in 1985, the prospect of the barrens being paved over moved from the hypothetical to the plausible. Still, the lands of the pine bush remained relatively open. It continued to serve as a de facto landfill, an exurban and biotic dumping ground. Like vacant lots in cities, it attracted random fires; trash fires became part of its pyric mix. Some 65 large fires burned between 1935 and 1987.[6]

This was a pittance compared to the routine, landscape-scale burning that had shaped the APB over thousands of years. By the 1960s a new, more fire-free order was imposing itself. Flames flickered like a guttering candle. Fire leached away. Fire protection succeeded, and fire-catalyzed biotas faltered. Without the pruning of open flame, pitch pine thronged on formerly plowed fields, oak thickened, aspen and invasive hardwoods like maple and cherry moved in. The pine barrens, long fabled as open and sunny, became impenetrable, a thicket of ecological gloom. A once-dappled scene of shrubs and grasses and flowers, a barrens savanna stocked with scattered pitch pine, was smothered under a tangle of woody plants and litter. Rare species became rarer. The Karner blue butterfly teetered on the edge of extinction. The pine bush was rapidly evolving into something else, which left the biota slowly strangling. It needed ecological shock therapy. It needed fire.

But first it needed a controlling agency with a cause. A grassroots campaign emerged, and in 1973 some 472 acres, called the Pine Bush Unique Area, was set jointly aside by Albany and Guilderland and the state Department of Environment and Conservation. Later, an additional 1,500 acres were added. The organizers recognized that they needed better information about the ecology of the pine bush and realized that fire was not simply a nuisance but a necessity. In 1986 they contacted the New York Field Office of the Nature Conservancy to assess the preserve's fire scene and possible operations to maintain it. The City of Albany, as part of a generic environmental impact statement (forced by the proposed expansion of the landfill), commissioned several scientific studies that addressed basic questions of fire history and habitat requirements. At the time it was not known what fire regime was suitable. For that matter, it was not known what minimum size and mosaic was required to sustain the biota, particularly the Karner blue butterfly.

In 1988 the New York State legislature created the Albany Pine Bush Commission as a public benefit corporation to coordinate action among the many owners within the preserve and the multiplying shareholders outside it. The commission includes representatives from the three municipalities, the state Office of Parks, Recreation, and Historic Preservation and Department of Environmental Conservation, Albany County, the Nature Conservancy, and four members appointed by the governor.

The commission was state funded through a real estate transaction tax, but not itself a state agency. Its charge was to preserve the pine bush habitat. The APB held 20 percent of all the species of greatest conservation need listed for New York.

Inevitably, scientists argued over particulars of data and interpretation. But a consensus developed (and thanks to a lawsuit became a legal mandate) that the preserve required at least 2,000 acres—more would be better—to retain ecological integrity; and those lands would have to burn. The TNC report, issued in 1991, made the case that the commission should, as the classic Latin phrase put it, *festina lente*—hasten slowly. A solid program needed a lot of test fires. It needed time to adapt to actual outcomes, not forecast ones. It would appreciate that fire could only do its work if the land was in a form that would allow fire to be controlled (a suitable fuel complex) while encouraging fire's full-spectrum catalytic effects (an adequate biotic matrix). All this was new. No one had tried to restore fire to the region's barrens and pine bush. No reference point existed that could serve as a beacon and index. They would have to invent the prescriptions and protocol. They took nearly 10 years before they launched serious operations. Hasten slowly.

By 2017 the preserve held 3,300 acres, gerrymandered through pine bush and suburb. They got some extra land as a trade-off for expanding the Albany dump. They got some more along the right-of-way of high-tension power lines belonging to National Grid (the power company did the first, heavy clearing). The preserve reckons it needs at least 5,300 acres to stabilize the scene. Of these 200 (6 percent) are burned annually. But it's in the burning that the story gets interesting.

———————

Fire is what its setting makes it. To get the right kind of fire the Albany Pine Bush Preserve has to fashion the right context. The combinations of treatments, and their sequencing, along with the timing and variations of firing, can resemble a Rubik's Cube in their complexity. Some of those "treatments" involve radical surgery. Restoring can mean rebuilding from the soil up (a demo patch from what was once a parking lot exists outside headquarters). It's ecological tough love.

The patches that make up the preserve vary in their deviation from historic norms. The ideal biomes are swathes of grassy savanna (rife with blue lupine) within the clusters of pitch pine-scrub oak savanna (rich with shrubs and forbs). Getting there can mean cutting and culling, drilling and filling, mowing and masticating, sowing and burning. Managers thin out pitch pine where it inhibits savanna and poses threats of crown fire. They clear-cut invasives like black locust groves, which not only shade out lupine but alter the soil biology and so must be extirpated by their roots, every fragment. They drill into birch and cherry and fill with herbicides. They brush-mow scrub oak. They plant lupine. They burn.

They've learned that not all fires are equal. Each has its place and time. Collectively, they have rules of assembly. Begin with dormant season burns after mowing oak. That will kill the oak but not burn up the stems. Burn next during the growing season so the lush growth will dampen the fire as it consumes the previous kill. To prevent too-rapid conversion— leaving refugia—no patch can be burned next to a recently burned patch. To keep down smoke each patch must be mopped up by the end of the day. The small-unit patches create lots of perimeter to defend relative to area burned, aggravated where that perimeter is a strip mall or office park.

The social preparations can be as demanding. Each burn requires 50 to 100 phone calls and involves mailing a ream of postcards. Managers answer to a commission of landowners and political appointees. Adequate staffing requires crews assembled from Department of Environment and Conservation rangers, town fire departments, the Nature Conservancy, the U.S. Fish and Wildlife Service, and the National Park Service. Taking a landscape, however distorted it might seem to ecologists, "back to the Pleistocene," can arouse passions, especially among those who have grown up with the old pine bush and assume that the landscape of their childhood is the true and only state of nature.

The upshot is that a place that had no reference point has become, for the Northeast, the model for getting fire back into places that need it. It serves as the flagship for the New York State Fire Initiative, a metropole for repopulating colonies of Karner blues elsewhere, and the standard for restoring similar pitch pine-scrub oak sites like the Montague Plains in central Massachusetts. It can also stand as an object lesson for the rest of the country.

Really?

To fire officers accustomed to wildlands, not countryside, who calculate scaling in the tens of thousands of acres, who have few opportunities to prep burn plots beyond cleaning fuelbreaks, who backfire on an order of magnitude larger than the whole APB Preserve, the fire program at Albany Pine Bush can appear negligible. This doesn't feel like genuine fire ecology. It doesn't look like restoration burning. It looks like the wildland equivalent of colonial Williamsburg. Besides, APB sends its fire techs out West for experience, western fire staff don't cycle into APB.

They're wrong. A true fire ecology of Earth must include arable fields, woods pastures, and transhumant grazing regimes, as well as back-of-beyond wildlands. It must include the cultural heathlands of northern Europe and forest plantations in Brazil. It must include broken biotas in need of restoration. And it needs to include pine bush preserves. All landscapes are now, in some degree, cultural. You can't understand fire's ecology until you understand it in all its settings.

What the Albany Pine Bush contributes especially is to remind us that fire is deeply biological. Knowing fuel loads is useful for predicting fire behavior, but not, in such circumstances, how those fuel loads happen and what the effects of fire behavior are. They don't answer the questions about sustaining habitat; they don't say how to keep Karner blues prosperous. The biota doesn't exist within a matrix of fuels: the fuels exist within a biotic matrix. Fire alone won't drive out black locust or promote lupine, whatever the fuel loading. Set properly, it acts as an all-spectrum catalyst. The load of mycorrhizal fungi in the soil matters as much as the load of 10-hour fuels on the surface. Fire is an accelerant: the driptorch is a fulcrum to leverage other treatments: it interacts with its ecological medium. Albany Pine Bush burns because fire does biological work that nothing else can.

What about getting burning to scale? Scaling fire practices doesn't only mean scaling up. It can mean scaling down. At some point, up and down, the results are not simply more or less of the same, but a change in kind. It means a toggle switch, not a rheostat. It means knowing how to reconfigure burn plans to satisfy the tiny. As America's estate continues to fragment, small-scale burning under tight constraints will become more of a norm. The Albany Pine Bush Preserve shows how it can be done.

The preserve expands our notions of what kind of institutions can govern landscape fire management. A solid program needs stable funding, political attention without political meddling, and hybrid agencies and collaborations. It needs a clear mission statement that can be made operational. It needs a fire culture, however eccentric it might seem to someone from the San Gabriels or the boreal realms of the Yukon Valley. The Albany Pine Bush shows what the Northeast requires for a successful fire program.

Paradoxically, it may be easier to transfer experience from the East to the West than from West to East. If it can work in the Northeast, it can probably work anywhere. It might even work in that confusing era that we call the Anthropocene, a future that won't be restricted to relict patches, however large, left from the Pleistocene.

WHERE YOU FIND IT

Staten Island

FIRE, LIKE RATS, can flourish wherever people and conditions favor it. It can burn in vacant lots as well as wilderness. It can burn within cities as well as along their shrub and conifer fringes. It can burn in long-settled places as well as newly unsettled landscapes. It can burn, routinely, on Staten Island.

Giovanni da Verrazzano reportedly first sensed the New Jersey coastline by smelling smoke. The earliest Dutch settlers observed routine fires through the meadows and scrub oak and pine of New Netherland. The sandy coastal plains of the Northeast are, in fact, the region's premier firescape—patches that dapple the interior as barrens and savannas carry those flames inland, popping up like gopher mounds. In the past these were far more extensive than they are today. As the land was plowed or paved or overgrew with mesic woods, their fires shriveled. Other than the New Jersey Pine Barrens the fires attracted little beyond local attention. An exception might be those on Staten Island, which seem an alloy of the novel with the archaic, a sense that a welcome native, long gone, has returned.

━━━

In truth, the fires never left. They have burned as far back as we have records of settlement. But surely, as New York City grew into Gotham, the nation's largest metropolis, its fires had vanished into an obscure past in which Wall Street actually ran alongside the wooden palisade of north

New Amsterdam and Washington Irving could write a mock epic of Peter Stuyvesant. Not so. So long as the land remained open, so long as "brush" and marshy sedges persisted, so has fire.

Unlike the meticulously cultivated Central Park, the ruder barrens of the outer boroughs could rub residents and rough into fire when conditions permitted. The potential for fire persisted like spores awaiting the right circumstances. Or more aptly, perhaps, like a pyric version of avian flu, the flames, usually no more than a seasonal inconvenience, periodically mutate into more virulent forms that flash through the brush from time to time. Still, watching landscape fires burn within the city limits is like finding a moose in Times Square.

The *New York Times* conveniently chronicles the evolution. In 1892 surface fires were apparently routine, like trash burning. The public's concern lay with "underground fires which are peculiarly hard to fight. They have thus far proved exceedingly annoying, and may soon endanger property." Surface fires replaced them, and they were nearly annual, rising in alarms as development put more lives and property at risk. In 1923 fires burned at both ends of the island, north and south, with a million dollars in damages. The next year 5,000 motorists watched firemen battle blazes to keep them out of a string of summer cottages. In the 1920s Staten Island experienced roughly 2,000 brush fires a year, or an average of nearly seven a day, though most clustered in the spring before green up and after fall dormancy when fuels were driest. Outbreaks of 35 to 50 in a day were news, but not rare. The fires of May 5, 1930, razed 82 bungalows at Oakwood Beach, damaged 200 others, forced thousands to evacuate, and prompted a rare "master call" (greater than a five-alarm) that dispatched 20 fire companies from Manhattan and Brooklyn. Other papers breathlessly reported that "New York City was encircled by brush fires."[1]

Railroads, rubbish burning, "pleasure seekers," and arsonists were common causes, though over the years trains and trash shrank in significance. What primarily remained were boys in the woods. The timing of ignition clustered around noon to 1 p.m., and after 3 p.m., which is to say, when school was not in session. They were a collective threat "The unwritten custom is that every one must join to fight a forest fire," the *Times* editorialist intoned, "whether he be a resident or only a transient. At best it is a thankless task. During the last few days the effort appears to have been largely futile."[2]

The May 1930 fires raged across New Jersey, Pennsylvania, and Rhode Island, as well as New York. Governor Franklin Roosevelt closed the state's forests. An editorial likened the scene to that typical of California. More fires came in the years to follow. During saturation days, other boroughs sent engine companies to assist; 1943 and 1947 had multiple days with outbreaks—the 1947 irruption got lost in the chain of bad fires that finally blew up in Maine. In 1955 Staten Island got a special Brush Fire Patrol. Then came April 1963, the year of record. This time the New Jersey Pineland fires claimed the public's attention, though the slopover into Staten Island and Long Island got plenty of ink.

Over the years, however, in the city as throughout New England, as land use intensified, fire protection improved, and people preferred internal combustion to open flame, fires retreated into ashy memory or protected pine barrens, the vacant lots of regional geography. The nation's fire problem went West. Landscape fire, especially wildfire, was something across the Great Divide of the Hudson River, like bison and brown bears. Or so it seemed.

Through the 1970s, the 1980s, the 1990s, the fires endured as background radiation, with occasional small outbreaks, like measles in pockets of unvaccinated populations. But even as the city pushed for stronger fire protection, it was reserving land as parks that would retain a fire potential. Then, on September 11, 2001, New York suffered the worst fatalities from urban fires in American history. Terrorists had chosen Manhattan's Twin Towers rather than Staten Island's brush and marshlands, but New Yorkers became newly resensitized to open flame in the city.

On March 15, 2006, the National Weather Service issued red flag warnings for New York City and surroundings. Historian Peter Charles Hoffer, with *Seven Fires: The Urban Infernos that Reshaped America* in press, wrote an op-ed for the *Times* asking whether the New York Fire Department was really ready for a major brush fire. His referent was the 1991 Tunnel fire that ravaged Oakland (the California allusion, again) which had humbled urban fire strategy by behaving like a wildland fire. Parks and open spaces that had once served as urban fuelbreaks could now act like fuses. With arson, global warming, and high winds "the smallest brush fire can become a city-devouring inferno."[3]

The annual roll call of burns continued. In 2010 fires burned for three days through reeds at Fresh Kills landfill, subsequently absorbed into the

Great Kills Park, a portion of Gateway National Recreational Area under the administration of the National Park Service. In 2012 a five-alarm fire at night burned again in Great Kills. The formula was eerily similar to that afflicting the West: invasive grasses (phragmites), dead-end streets, cityscapes abutting firescapes, ample ignition. Gateway has the fourth highest incidence of wildfires in the NPS. Fuel management plans are underway for mowing, with possible prescribed fire, and longer-range ambitions to eradicate the reeds, replacing them with still-flammable but less volatile species.[4]

Fire is where you find it. It can burn across mountains or around town-houses. Staten Island is not unlike the rest of New England, mostly built up, mostly controlled by people, but vulnerable to fire where it isn't. Fire is tucked away in niches. It's where you don't expect to find it. In some places it just is; in some it's where it needs to be. In Staten Island it's a curiosity, like a biotic Occupy Wall Street in Zuccotti Park. It could be eradicated if the city decided it wanted to abolish the open space that houses it. The city has fires because it wants that open space, so the problem appears as more a threat and an occasional scare than the flash point for a riot. But it might not take much to cross the threshold.

THE FOREST AS GARDEN

Charles Sprague Sargent

C HARLES SPRAGUE SARGENT was a botanist, a Harvard professor, and, as director of the Arnold Arboretum and a man with private wealth, a latter-day Boston Brahmin. He didn't wander the West in search of enlightenment, adventure, and progressive politics. He was as likely to look to Europe or the floral splendor of Japan for inspiration as to the Rocky Mountains. He lectured by doing rather than exhorting. He grew his ideas out of science. He didn't found a movement. He didn't inspire a corps of acolytes. A member of the Establishment, he worked within the system. He stayed resolutely in a middle vision of what America's forests might be, hovering in a kind of Lagrange point between the gravitation pull of his better known contemporaries, Gifford Pinchot and John Muir.

His social standing, his long affiliation with Harvard's arboretum, his temperament, his career—all mark him as an odd character to influence national conservation, much less national fire policy. Yet he did. More than once.

Charles Sprague Sargent was born into privilege—the son of a wealthy merchant in the East India trade, of a long-standing family, educated in private schools, graduated from Harvard in 1862. But like many of his caste, he carried a sense of social obligation as well. He enlisted in the Union Army, a lieutenant in the Second Louisiana Infantry, then aide-de-camp for the Department of the Gulf at New Orleans, and brevet major of volunteers for "faithful and meritorious service" during the campaign at Mobile. He mustered out in August 1865 and promptly toured Europe for three years. By the time he returned to Boston his interests gravitated toward botany and horticulture, and especially their fusion in gardens. It was an ideal blend of science, practice, and aesthetics. Returning to Harvard, he became a professor of horticulture, director of the Harvard Botanic Garden, and in 1873 director of the newly conceived Arnold Arboretum, subsequently appointed the Arnold Professor of Arboriculture. Two days later, he married Mary Allen Robson. The remainder of his life built steadily from those events.[1]

He was, a memoirist wrote, "built for the long stride." That was as true for his life as for his vigorous excursions to gather specimens. He combined a collector's zeal with a scientist's patience. He moved with a deep conservatism that made it difficult to "swerve" him from his determined course, or to "retard or accelerate" that progress. His routine, "quiet, never hurried . . . forging ahead with irresistible momentum," yielded an astonishing corpus of work. There were excursions that made him a latter-day American apostle of Linnaeus—the Northern Pacific Transcontinental Survey (1882–83), to the forested regions of the United States in connection with the Tenth Census, to the Adirondacks, the Cascades, and the Rockies, to the Caribbean islands and the Bahamas, to Mexico, to Japan, and in 1903 a tour around the world, including the Crimea and the Caucasus and a trek across Eurasia on the Trans-Siberian Railway before visiting China, Singapore, Java, and Japan, all the while collecting seeds from trees and shrubs. In 1905–6 with his son Robeson he collected in Peru and Chile, then the Falkland Islands, Brazil, the Cape Verde Islands, and Portugal.[2]

That gathering of specimens and experiences he made public in two venues. One took the form of publication, of which he created an endless bibliography, culminating in a series of summae, any one of which would have justified a scholarly career: *Report on the Forests of North America (Exclusive of Mexico)* for the Tenth Census; a 14-volume *Silva of North America*; and the *Manual of the Trees of North America (Exclusive of Mexico)*. In 1895 he was elected to the National Academy of Sciences. His formal herbaria he translated into a widely influential journal, *Garden and Forest*, and near the end of his life, *Home Acres*, which summarized his preferences. These publications nurtured his collections into print. The other outlet was the Arnold Arboretum, where he cultivated the plants themselves. Here he translated raw nature into an ideal landscape, what he regarded as "the greatest garden in America." More than a living herbarium, a kind of Kew Gardens for the American empire, it advertised a model relationship between people and nature. It was a task for the long term, perfect for a man of the long stride. The endowment for the arboretum was inadequate, there was no infrastructure in the form of buildings and library, and the land itself was a "worn-out farm partly covered with natural plantations of native trees nearly ruined by excessive pasturage." It was, after a fashion, a metaphor for much of depopulated New England.[3]

Sargent began its rehabilitation with careful planning, enlisting Frederick Law Olmsted, the architect of Central Park in New York City and advisor on Yosemite Valley during its brief stint as a state park. What emerged was neither a pristine preserve nor a commercial plantation, but a working landscape for botany and public education, a kind of "great out-of-door museum" and part of a linked chain of parks in Boston. If Yellowstone seemed to represent prelapsarian Nature, so a Garden, in Sargent's mind, could rebuild what had fallen. Unsurprisingly, Sargent became an advocate for public parks, nature's own arboretums, working in the Adirondacks and later campaigning for what became Glacier National Park.[4]

In brief, his scientific career was outstanding, his social standing impeccable, and his judgment sought on matters of importance. He entered American fire history when he chaired committees that shaped state-sponsored conservation.

———

He had a patrician's sense of duty. He served as president of the Massachusetts Society for the Promotion of Agriculture; the vice-president of the Massachusetts Horticultural Society; a trustee with the Boston Museum of Fine Arts and the Brookline Library; and park commissioner for Brookline. His national service came with overseeing the forest survey for the Tenth Census; with chairing the Commission on the Adirondacks;

and with chairing (again) the National Academy of Sciences Forest Commission.

In his 610-page *Report on the Forests of North America* for the 1880 census, Sargent methodically documented the character of America's wooded estate and the depredations that threatened it. The *Report* included the country's first census of fires—Franklin Hough's contemporaneous survey was more anecdotal—and its first exercise in pyrocartography at a national scale. What John Wesley Powell had isolated in the disturbed slopes of the High Plateaus of Utah, Sargent had broadened from sea to sea. America, it seems, was the Brazil of its day, burning for all and any reasons; and intellectuals then, as today, condemned what seemed to them adolescent vandalism.

They regarded laissez-faire burning to be as devastating as laissez-faire logging. "The extent of the loss which the country sustains every year from injury to woodlands by fire is enormous." He elaborated before the Massachusetts Board of Agriculture that "could the actual amount of such losses be computed they would astonish even those most familiar with the condition of the American forests. . . . The extent of forest fires throughout the country is infinitely greater than has ever been seriously supposed." The loss of timber was only the beginning. Worse was the destruction fire inflicted on the soil and watershed, and on "the confidence of the community in the value and stability of forest property." Without fire protection no one would invest in forest land, and neither would the country. A prolonged discussion ended with the moderator announcing that "this discussion has gone as far as is profitable without any direct action." With Sargent's assistance the board would again work with the legislature to strengthen laws to discourage wanton burning.[5]

He took his conclusions to the country with an 1882 article in *North American Review*. "Forest preservation, as a national question, must soon occupy public attention. The problem involved is one of grave import, and its solution is not easy and cannot be immediate. . . . The American people are still ignorant, not only of what a forest is, but of the actual condition of their own forests, and of the dangers which threaten them." Of the Northeast, "its wealth has been lavished with an unsparing hand; it has been wantonly and stupidly cut, as if its resources were endless; what has not been sacrificed to the ax has been allowed to perish by fire." He then introduced a term that would become, in the hands of Progressive reforms like Gifford Pinchot, a rallying cry: "timber-famine."[6]

With his reputation as the nation's foremost arborist confirmed, Sargent was asked to chair a forest commission established by the State of New York to report on forest protection in the Adirondacks. Its report noted both the failure of agriculture and of forest industry. "Farms are cleared, two, or at most three, meager crops are snatched from the cold stony land; and then starvation drives the settler, exhausted in the fruitless struggle with the uncompromising and unforgiving fate of nature, to abandon his fields dearly purchased at the price of indescribable suffering and privation, and seek a new home, which in turn must be abandoned at the end of a few years." Equally, slash-and-burn forestry would fail to provide a sustainable economy and would trash the watersheds that were so vital to commerce. At present, Sargent concluded, the damage was not horrific, but there was nowhere in the mountains that the axe could not penetrate and fires not inevitably follow, from half a dozen sources. "Fires are slowly and surely destroying the Adirondack forests." They were moving from the outside in, and "unless they can be stopped or greatly reduced in number and violence, nothing can prevent the entire extermination of these forests." Only protected forests, managed against the depredations of axe and fire, could stem the prospect.[7]

That experience turned out to be a trial run for the 1896 National Forest Commission established by the National Academy of Sciences to report on the country's fledgling forest reserves. Again, Sargent chaired—had been, in fact, one of the small band who helped formulated the plan. The commission's other members were General Henry L. Abbot (retired from the Corps of Engineers), Alexander Agassiz (Harvard Museum of Comparative Anatomy), William Brewer (Yale and state botanist of California), Arnold Hague (USGS), Wolcott Gibbs (NAS, ex-officio), and the young Gifford Pinchot. Later, when Gibbs and Agassiz declined to join its tour of western forests, John Muir did.

The commission was charged with three questions, the first of which was "Is it desirable and practicable to preserve from fire and to maintain permanently as forest lands those portions of the public domain now bearing wood growth for the supply of timber?" The second question expanded interest from timber to forest influences generally. The third addressed future legislation. But the first, founding question was fire protection.[8]

Reserves and protection were the twin pillars of forest conservation globally. What happened in the United States evolved in parallel with what Britain, France, and the Netherlands were doing in their colonies. The first question the commission posed to itself is identical to the founding question raised by the Indian Department of Forestry in 1878, whether it was possible to control fire, and if possible, whether it was desirable. The American response was similar to that taken in colonies and settler societies elsewhere. "It is practicable to reduce the number and restrict the ravages of forest fires in the Western States and Territories, provided details from the Army of the United States are used for this purpose permanently, or until a body of trained forest guards or rangers can be organized." The first task was to secure the forests from depredations.[9]

Within that grand global strategy there was plenty of room for difference. The forest commission disagreed on what the forest reserves meant and who should administer them. The split was most pronounced between Sargent and Pinchot. Pinchot saw the forests through the eyes of a trained forester, who wanted a civilian corps of foresters to oversee them. Sargent viewed them with the vision of a botanist and creator of parks, who believed that, at least for the present, the reserves should fall to the army, as the national parks had since 1886. The chief threats were trespass and fire. His view is what the commission recommended, though the Organic Act of 1897 left administration with the Department of the Interior.

Sargent was not opposed to commercial timber—had in fact argued that fire protection in Massachusetts (and elsewhere) was required to sustain the industry. "The forest question has become a question of dollars and cents. We can no longer afford to allow our forests to burn." But he saw the forest reserves more as parks than as plantations. They were nature's grand arboretums. Sargent rallied Muir to his cause. The fissure between Sargent and Pinchot, as that between Pinchot and Muir, widened over the years.[10]

Behind this stance lay the Arnold Arboretum and *Garden and Forest*, both of which promoted a more ancient vision of a park as a garden, and a garden as a pleasuring ground. Eden had been neither a woodlot nor a wilderness but a garden. As his memorialist put it, out of Sargent's life's work "has emerged a national policy of intelligent forest conservation and utilization, of salvaging the relics of lumbering, and of preserving for future generations samples of Nature's own great arboretum in the form of national parks."[11]

Behind that vision, too, lay an understanding of landscape history and responsibility. It was a vision based on protecting what was unsavaged and of renewing what had been wrecked—a dyed-in-the-wool New Englander's view, though one informed by a world traveler and world-class scientist who had painstakingly rebuilt a rundown farm into the forest gallery that became the Arnold Arboretum. His labors at the arboretum were to its day what Aldo Leopold's efforts at his Sand County farm were a half century later.

Charles Sargent's vision did not become the national norm. The institution that eventually emerged from the work of the forest commission, notably the U.S. Forest Service, took on its own life, fashioned in the image of global forestry, and then after the Great Fires of 1910, remade fire policy in its own likeness. Yet it is worth recalling that the ideas that Sargent gave substance to occurred amid America's first Gilded Age, when capitalism had an unbridled hand and government's first duty was to do nothing. The triumph of Pinchot occurred later, when the country rallied to the call of Progressive reform.

Within the national panorama the New England landscape increasingly seemed less an archetype than an anomaly. Its 281 acres, even as part of Boston's Emerald Necklace of parks, seemed laughably quaint when compared to the Bitterroots or the wooded swell of the Cascades. Yet Sargent's view fit New England far better than those of Pinchot or Muir. This was an unblinkingly cultivated landscape, more recently a feral one because its cultivators had moved West or into cities. The task was to rebuild, and then to export the best of that hard-won wisdom to those lands of the national estate not yet slashed and burned. The issue for the future was whether its forests would regrow with artistry or with abandon.

Charles Sargent did not find himself among the pantheon of America's fire gods. It's certain he never held a pulaski. He was more familiar with a trowel than a shovel. He probably burned nothing more than autumn leaves and the occasional steak. But he bolstered state-sponsored conservation at a time when the nation desperately needed it, and when the forces for asset-stripping the nation's natural wealth were never more powerful. Along with fellow New Englander George Perkins Marsh, he made the case for protection and preservation, and unlike most others, he translated his ideals into practical measures, working within the system to move it in new directions. He put numbers to the national outcry over untrammeled burning. He proposed legislation to contain burning. He crafted institutions to apply reforms. He patiently built a material alternative that could show the public what could be done.

In all that, and in his forgotten stature within fire history, he can stand for New England overall.

PITCH PINE AND LEAST TERN

AKE ALMOST ANY MEASURE of New England, and Massachusetts sits close to the core. Geography, history, culture—the Bay State held the most vigorous colonies, led the revolution, developed the most robust institutions of national import. New York eventually outpaced it, but Boston remains the great metropolis of New England. Maine, while spun off from it in 1819, held more land, and more unorganized lands, but Massachusetts developed most of the ideas and institutions that governed New England's land history. Early American art and literature largely emerged here, as did the antebellum American Renaissance. The industrial revolution made landfall along its fall line. The first formal scientists in America joined the faculty at Harvard and Yale. Massachusetts commissioned the first geological and botanical surveys by an American state, a model for other states and subsequent federal surveys.

So it is no surprise that Massachusetts also holds special standing in Northeastern fire history. The historic fires occurred in the big states, Maine and New York, and its giant pine barrens have granted New Jersey pride of place for explosive burns, but Massachusetts has occasionally burned with proportional dimensions, and littoral Massachusetts has rivaled the New Jersey Pinelands on occasion. From time to time Cape Cod might have been better named Cape Flame.[1]

Its landscape fires were a novelty of natural history for early chroniclers. "The Savages are accustomed," explained Thomas Morton in *New England Canaan* (1637), "to set fire of the Country in all places where they come; and to burn it, twize a year, vixe at the Spring, and the fall of the leafe." Like other chroniclers Morton did not say the indigenes burned everything everywhere. The accounts make it clear that the natives burned where they traveled and where they had particular places of interest, such as swidden fields, hunting grounds, and around villages to protect against wildfire. Chronically wet lands didn't burn except in droughts. (Morton noted that if you want "good tymber," you had to seek for them "in the lower grounds where the grounds are wett when the Country is fired.")[2]

Places without recurring interest didn't burn, or because they couldn't burn held little interest. But there was far more fire than would appear possible from today's perspective, where boutique burning has become the norm. Naturalists described open fields ("as in our parkes") that today are thronged with woods (and houses). They speak of pine and oak as dominant throughout the drier countryside of southern New England. Many of these practices the newcomers from Europe took over, and then added others peculiar to herding, potash manufacture, charcoaling, and changes in hunting habits, in addition to pushing agriculture, with the leverage of the axe, into areas previously immune to fire. Eventually, the fires were either domesticated into cultivated fields and pastures, or banished by stripping away their fuel stocks.

Thereafter Massachusetts fire history surfed the crests and troughs of New England settlement history, as land clearing and land reclamation shifted human fire practices, and from time to time littered the land with feral fuels. For the 1880 census C. S. Sargent summarized both the long-wave history of the woods and their current flammability. "The original forest which once covered these states has disappeared and been replaced by a second, and sometimes by a third and fourth growth of the trees of the Northern Pine Belt." In places the regrowth barely kept pace with continuing demands, particularly for fuelwood. But the prognosis was good. "Abandoned farming land, if protected from fire and browsing animals, is now very generally, except in the immediate vicinity of the coast, soon covered with a vigorous growth of white pine." That regeneration promised "to give in the future more than local importance to the forests of this region." Certainly the slash from the reclearings contributed to the regional fire load, which influenced national thinking. In 1880, Sargent noted that the southern New England states sustained "a considerable annual loss from forest fires. In Massachusetts that year 13,899 acres of woodland were reported destroyed by fire, with a loss of $102,262. Of these fires fifty-two were set by locomotives, forty by fires started on farms and escaping to the forest, thirty-seven by hunters, nineteen by the careless use of tobacco, eight through malice, and three by carelessness in the manufacture of charcoal."[3]

The scene worsened as the old-field white pine was logged and its slash fired. But then the new order established its grip: the region completed the pyric transition from burning living biomass to burning lithic biomass, though given its rejuvenating forests fuelwood usage remained proportionally high. When fire protection became a political issue, the states created forestry bureaus to assist local communities. Open flame subsided from the land.

Over the past century, Massachusetts's fire history could represent the region's. Synecdoche, letting a part stand for the whole—a term from ancient Greek rhetoric that Ralph Waldo Emerson or Henry Thoreau would have known—aptly characterizes Massachusetts's place within New England fire history. Through the 1920s the slovenly logging of old-field white pine left slash strewn widely; fires followed. In 1927 alone some 16,000 acres burned at Townsend State Forest, 7,000 acres ripped from Erving to Wendell, and a fire blasted out of the Montague Plains and destroyed the village of Lake Pleasant. Then exhaustion combined with Depression to crash the market. A few fires bubbled up here and there, particularly in the sandy pine-oak southeast. Then the 1938 hurricane trashed a significant fraction of what remained, and that debris was never wholly cleaned up before war ended the project (and the CCC).

The contemporary era begins with burns in 1946 and 1947, the latter an outlier of the Maine conflagrations. Then came the 1957 Plymouth blowup (12,000 acres), followed by an echo in 1964 that swept over 5,500 acres and incinerated 26 buildings—Massachusetts's contribution to the series of drought-fed Sixties' burns that moved along the littoral from New Jersey north. Almost nothing comparable has succeeded those outbursts. Rather, each paired decade after the 1910s–1920s dropped in area burned. The 1930s and 1940s were half as extensive as the two before it (but high enough to qualify for two dioramas in the Harvard Forest suite). The 1950s and 1960s were again almost half as large. By 2000 fires burned a twentieth as much as a century before. A graph shows waves and troughs, each crest lower than its predecessor. It's the classic curve of a dampener.[4]

The reasons are many, and the fact that there are so many speaks eloquently to the complex, nuanced character of fire in New England. The big three explanations cluster around land use, better control over ignitions, and far superior suppression capacity. Rural lands were being replaced by suburbs and exurbs, which brought a change in fire habits and the availability of fuels. Those lands were no longer felled as before, portable sawmills left, and what logging did occur bequeathed less slash. Grasslands yielded to woods, especially fire-intolerant hardwoods. Scrub grew into mature forests, less prone to explosive burning. Where conifer reproduction and berry patches persisted, they were typically overtopped by canopies that shaded the surface. Meanwhile, fewer people burned trash, debris, and fields. There were also fewer locomotives littering sparks; fire bans and burning permits were enforced; and Smokey Bear found an enthusiastic audience. Extensive roads fragmented blocks of combustibles and allowed access by crews and engines. A permit system for burning, mandated seasons for burning, and outright fire bans reduced ignitions. Volunteer fire brigades learned to cope with wildland fire. A gaggle of fire wardens evolved into a system. And locals turned to the state for assistance. The state had, in turn, created forests and parks, complete with the apparatus for protection, and what the state couldn't provide, it could get through mutual aid agreements, especially the Northeastern Forest Fire Protection Compact (NFFPC). Local fire chiefs remain—this is the heartland of home rule, after all—the responsible authority. The state, through the Bureau of Forest Fire Control, nestled within the Department of Conservation and Reservation, gives those 353 chiefs, many of them heading volunteer departments, extended capacity and connection that they can't muster on their own.

In brief, the land evolved beyond its extreme fire-proneness, more fires were prevented, and fires that broke out were caught by aggressive suppression. These are traditional methods, but they work here, as throughout New England, because there is scant natural fire to push in if anthropogenic fire pulls out. Still, the traditional goal of American forestry was to reduce fire loss to 0.1 percent of land area under protection. Massachusetts weighs in at 0.12 percent. There are a lot of fires, many in the 5 to 50 acre category, which add up. Federal agencies tend to have a lot of small fires and a few large ones (less than 3 percent) that account for the area burned. Massachusetts still has a healthy percentage of midrange burns.[5]

Of course Massachusetts did all this with its distinctive accent. It erected a fire tower network early, and kept it when others scrapped towers in favor of aircraft. The 42 lookouts remain popular with both the public and local fire chiefs. It developed brush breaker engines, beginning with the aftermath of the 1938 hurricane, and then evolving them to adapt to the military vehicles acquired under the federal excess equipment program, sometimes using designs devised at the Roscommon Equipment Center, often with local innovations. They serve as armored divisions, an alternative to bulldozers, capable of pushing over 8- to 10-inch diameter pitch pine and scrub oak. (They also serve as fire engines for flanking fires with wet lines, which are adequate in spring burns.) And Massachusetts must, in fire as in politics, deal with the exceptional powers of local jurisdictions, particularly townships. Like the biota, it's a fine-grained mosaic. The state supports the towns, and the compact and feds support the state.

———

Statistics at this level can be squirrelly. Massachusetts complicates the issue by reporting fires on a fiscal year, while everyone else reports on a calendar year. But even without adjustments Massachusetts has more fires and burned area than its neighbors. This, however, also appears to be an artifact of reporting: its still-extant fire lookout system records smokes that local authorities would not otherwise report to the state.

So it is not clear that the southeast quadrant of the state, the sandy pitch pine and scrub oak landscapes that thin and curl into Cape Cod, is more active than elsewhere or whether this, too, is a curiosity of the preserved evidence. Certainly the existence of public land in the form of Myles Standish State Forest, Camp Edwards, and Cape Cod National Seashore (formerly Camp Wellfleet, a training facility for tanks) lend themselves both to more fires and better reporting.

But whether real or apparent it's a hot spot. Since the late 19th century big fires—1,000 acres or more—came every 10 or 20 years; after 1964, the return interval lengthened to 30 to 50. The 1887 fires swept over 25,000 acres across four counties. In 1900 winds drove flames "over a great section of Plymouth County," with heavy losses, including "scores of frame buildings" and the life of Mrs. Joseph A. Brown, who "dropped dead from excitement" when the fire threatened her house. In 1923 10,000 acres burned in Bourne and Falmouth. In 1923 17,000 acres burned in Bourne and Sandwich. In 1930 16,600 acres burned in Barnstable. Some 5,000 acres burned in Sandwich in 1938. In 1941 fires springing out of dry marshlands incinerated 550 buildings in Marshfield. In 1946 between 5,000 and 50,000 acres burned at Camp Edwards (the records are confused), using debris left from the hurricane as an accelerant. Then came the 1947 fires. The 1957 Plymouth fires burned 15,000 acres in 12 hours; and 1964, with 51 fires, almost all incendiary, burned 5,500 acres and 26 structures. Pitch pine and scrub oak, often with an understory of ever-flammable huckleberry, were chronically aflame. Sea breezes complicated fire behavior. The 1937 fires, amid shifting winds, killed two firefighters during a burnover. The 1938 fires killed three. The 1971 fires burned over a crew of eight, injuring two severely. No wonder brush breakers became popular.[6]

Then the outbreaks abated. Camp Wellfleet became a national seashore, with tanks replaced by tourists. Camp Edwards substituted prescribed fire for wild fire. Rural countryside morphed into suburbs. Arson fires faded. Brush breakers strengthened. Lyme disease rather than flame haunted the locals.

===================

The state was cursed by its own success. Without recurring outbreaks, it's hard to press for the political support that maintains the level of protection. The fires were there, in abundance. They just no longer chewed through their shortened leashes. The problem for public lands became less about halting bad fires than reinstating good ones. The volatile pine-oak scrub burned easily because it was built to burn. Removing all fire had costs.

So, here and there, a patch, a park, a frost bottom, fires have been gingerly introduced. The basic argument is public safety—burning to reduce hazardous fuels and to train fire brigades, most of which are town departments. But the reason for fire rather than woodchippers is the ecological punch that free-burning flame brings and combustion in machines doesn't. It's a cautious project, feeling its way like a creeping ground fire through a public still besotted with Smokey Bear. Mostly, it occurs in islands, either isles surrounded by ocean or isles set apart by legal borders, or in some spectacular cases, both, all safely moated against breakouts and public impatience. Camp Edwards is a good example. So is the "fire triangle" nurtured by the Department of Conservation and Recreation in Myles Standish State Forest in what the agency hopes will be a public demo plot for promoting more expansive burning. The state wildlife agency supports burning on Montague Plain and Fall River State Forest.

The most striking is surely the successful program on Martha's Vineyard, begun as a preserve for the heath hen, later renamed as Manuel Correllus State Forest, then assisted by the Nature Conservancy. Here fire stimulates native flora and traditional grasslands vital to a slew of rare species, while fashioning a blackline fuelbreak against the wildfires that historically roar out of the southeast. (In 1930 one such fire burned 5,000 acres, and in 1946, another 5,120 acres.) High-end developments crowd the reserve's borders; Hops Farm has 240 homes valued at $158 million and Dodgers Hole, 526 homes worth $327 million, and more are planned. Smoke isn't welcome, but fuel abatement is. The trick is that the ecology needs fire, but the people don't, and they don't like smoke. There is little margin for error. All these are small projects, but not trivial, and not without both practical and symbolic significance.[7]

Perhaps the sweetest example comes from Lovell's Island in Boston Harbor. The indigenous least tern was being driven to extinction by a combination of exotic Norway rats and invasive grasses. Breeding pairs were introduced to the mainland, where they quickly became pests. The proposed solution was to restore the native Lovell's biota through fire. Here Department of Conservation and Recreation (DCR) crews crossed the harbor in boats and burned a half-acre swath along the shore. Three weeks later DCR wildlife biologists reported 93 least tern nesting sites on the recovering burn.

Half an acre? That would barely qualify as a pile burn in Florida or a test fire in Montana. Yet it is likely the future of Massachusetts (and New England) fire over the coming couple of decades. Hold the line against wildfire. Keep Smokey. Reintroduce fire in select places of high biotic value. Add least terns, Karner blue butterflies, pitch pine, and lupine. The effort only seems puny because the American fire community has traditionally fixated on vast public lands. Lovell's Island is closer to what most of the country actually looks like.

Massachusetts may have been a leader in the American Revolution, but it was a laggard in America's fire revolution, which had its anchor points in Florida and California. Now that revolutionary era has ended. The future will have to deal with mixed-ownership lands, with hybrid tactics that blur the boundaries between suppressed and prescribed fire, with managerial mashups among agencies. With intense urbanization in its northeast, fires prone in the southeast, dappled settlement and public lands to the west, Massachusetts eerily echoes the pyric geography of the country overall. The national future of fire will increasingly deal with big managed wildfires in the public domain of the West and with small landscapes, redolent with complex histories and competing goals. The future, or at least one likely version of the future, sounds a lot like Massachusetts.

THE WUI WITHIN

THE NATIONAL FIRE PROTECTION ASSOCIATION (NFPA) was incorporated in 1896, the year before Congress passed an Organic Act for the national forests. In 1905 the Forest Service took over those reserved forests; two years later the NFPA began publishing its journal. In 1910 the Big Blowup in the Northern Rockies traumatized the USFS and catalyzed a national program of fire suppression. The next year the Triangle Waist Co. fire similarly bonded urban fire protection to Progressivism.

As those chronological pairings suggest, the world of fire was fissioning into two realms. One was public wildland, overseen by foresters. The other was the built landscape of city and industry overseen by engineers and architects, and this was the domain of the NFPA. Over the coming decades each grew separately, buffered by a still large rural setting, until postwar sprawl swallowed up that countryside, binding those two fire realms with asphalt knots. Wildland fire overflowed into an urban fringe; structural fire protection scattered like embers into wildlands. The larger landscape displayed an increasingly fractal geometry of fire or, to shift metaphors, a non-Euclidean geometry in which seemingly parallel institutional lines might cross. The USFS had to cope with burning houses. The NFPA had to imagine codes and standards to govern fire protection amid wildlands.[1]

Still, those early tremors seemed peculiar to California, a fiery equivalent to its earthquakes. Even there no one outside wanted to own the issue. Instead, fire itself forced a fusion when flames in 1985 crashed through suburbs in Florida as well as California and consumed some 1,400 homes. Suddenly, the problem was big, national, and undeniable. Forest Service researchers, with western lands in mind, dubbed it the "wildland-urban interface fire problem." The next year the USFS partnered with NFPA to establish a National Wildland/Urban Interface Fire Program. Never had such a dramatic task begun with such a klutzy name.

The 1985 fires were the early signals of a long-wave drought that began to settle over the West. A whopping outbreak of fires struck California in 1987. The Yellowstone fires followed in 1988. When the ash settled, the wildland fire community was ready to turn from an obsession with wilderness fire to something seemingly less abstract and controversial. The WUI was the obvious candidate, and the National Fire Protection Association was the preferred partner. The NFPA could reach out to the urban fire services as the National Association of State Foresters could for wildland fire protection.

But the regrettably named WUI went beyond a fire of political convenience. More and more, the federal land agencies saw their mission distorted by the gravitational pull of houses burning along their borders. That telegenic spectacle deflected attention from fire's restoration, it redirected fireline strategies, it absorbed firefighting resources, it put fire costs on steroids. It threatened to turn fire as an integral part of land management into fire as another emergency service. And the problem got worse. The legacy of wildland fuels from long decades of fire exclusion, former rural lands now overgrown with houses and scrub, everything intensified by a stubborn shift to a drought-prone climate—all made the WUI into a black hole that threatened to suck everything else into its maelstrom.

The federal land agencies like the Forest Service tended to view the encroaching problem as an alien infestation, as though exurbs were an invasive exotic, though with an exoskeleton plated in wood. Such fires were neither in their mandate nor their experience nor their temperament. They were not equipped to fight them or to study them. The NFPA stepped into that gap. It brought its meticulous style of fire investigation to bear on the 1989 Black Tiger fire outside Boulder, Colorado, and in 1991 to the Spokane fires and the Oakland Hills fire. In 1993 the term *firewise* was coined. Both fire communities, urban and wildland, began to interact. The NFPA participated in National Wildfire Coordinating Group committees. The Forest Service sponsored research into how nominally "wildland" fires actually burned houses. The 1998 season with the International Crown Fire Modeling Experiment validated lab models, while the fires that year in Florida confirmed the urgency of the task. The national program, now known as Firewise Communities, scaled up.

The 2000 National Fire Plan brought serious money to the table. The next year saw a two-year pilot project underway, which then segued into a national program funneled through the state foresters. By now the program had many sponsors beyond the Forest Service, including the Federal Emergency Management Agency, the U.S. Fire Administration, the National Association of Fire Chiefs, and the National Association of State Fire Marshals, all with long-standing ties to urban fire services, while the Department of the Interior added to the wildland side. Within a decade Firewise had over 600 communities enrolled and was targeting a thousand. Those formally enrolled measured only a fraction of its influence as hundreds of others incorporated reforms on their own from the Firewise example.

The I-zone fire, as Californians called it, was a borderlands, a world of its own yet one that forced each of its paired sides to internalize the other. In 2010, recognizing that fire protection had shifted from urban cores to the fringe, the NFPA created a Division of Wildland Fire Operations. The next year, when the federal agencies published their National Cohesive Strategy for Wildland Fire Management, they identified one of their three primary concerns to be fire-adapted communities.

The understanding of the I-zone, however, continues to be shaped primarily by the wildland fire community. The Forest Service in particular identified the problem, named it, and funded programs to address it. The service saw the WUI not only as a legitimate fire issue but as a threat to its larger mission in land management and fire restoration.

It did not want to absorb the WUI so much as to hand it over to others so it could attend to its core purposes. It saw the outbreaks as private-sector fires that the private sector should handle. It turned to the NFPA to assist with structural fire as it did the Nature Conservancy to help with landscape restoration; Firewise Communities was the political equivalent of the Fire Learning Networks. These were national programs, not federal ones. At the same time the Forest Service handed over supervision of the National Incident Management System, which it had developed, to FEMA. It did not want to become an all-hazard emergency response operation: it was a land management agency. It regarded the WUI as an unwanted foundling abandoned on its doorstep.

Yet the NFPA offered more than a potential adoption agency. Its participation meant a chance to redefine the problem. The federal land agencies viewed sprawl as an encroachment, as a nasty problem that gnawed at its boundaries like a bark beetle infestation. They understood proper responses in terms of keeping that hazard at bay; the ideal solution was to simply zone exurbs out of the scene. It was equally possible, however, to view the issue from the other side of the border and define the WUI as a far-flung (if outrageous) extension of urban fire, as a species of city fires with peculiar landscaping. Until the early 20th century, American metropoli had routinely burned; then that melancholy chronicle of conflagrations stopped. How, exactly, had that happened? Might those same lessons extend into exurbs?

Viewing the WUI as urban fire could redirect the quest for remedies. It shifts focus away from wholesale landscaping and onto the structure itself, what became the "home ignition zone." It emphasizes exposure, the relationship between structures rather than between individual structures and their interstitial woods. It targets combustible roofing, open eaves, faulty glass windows, attic window mesh. It looks to questions of egress, or how firefighters might get in and civilians get out. It suggests that insurers would not be vital players. It emphasizes the voluntary adoption of model codes. It proffers a political as much as an engineering exemplar for protecting assets from unwanted fire.

Perhaps surprisingly, granted the urban fire services' passion to eliminate fire of any and all sorts, the NFPA understood that fire would happen. It did not seek to promote fire as a means of urban renewal as the USFS longed to reintroduce fire for ecological regeneration; but it accepted that fire was inevitable. In its own way it sought means to live with fire. For the Forest Service a fire-adapted community was one that could coexist with the wildland fires needed by its surroundings, that did not interfere with the agency's larger goals for land management; ideally it was one that was never built at all. For the NFPA a fire-adapted community was one that could survive wildfire. Its codes sought to eliminate as much of the hazard as possible. It scrutinized the house and its immediate, direct-contact landscaping. It regarded efforts to redirect or shut off the flow of sprawl as quixotic. Houses would happen. Houses would burn.

In 1981 the NFPA moved out of congested Boston to a campus-style complex in Quincy, adjacent to the Blue Hills, a prime patch of public open land acquired by the Metropolitan Parks Commission in 1893. As one might predict, it attended to landscaping as carefully as it did the interior workspace. The main building wrapped around a constructed pond framed by planted shrubs and trees. This was a place built to code, from its choice of window panes and door handles to its selection of flower-bed flora. It was also a case study in why the wildland-urban interface, or as some preferred, the American intermix, was so potent. No less august an institution than the National Fire Protection Association had planted its headquarters squarely in the I-zone.

There was not much threat from fire, not only because Boston is not an intrinsically fire-prone place but because the setting and structure had been meticulously sculpted to deny fire of any kind a presence. But the fact that the scene existed at all testified to the social drivers behind the WUI. This is where many Americans wished to live. They wanted to see nature, they wanted a buffer of privacy, they wanted a secure setting. Given a choice they would spread out rather than build up. The NFPA was no more or less than the society that sustained it. Its behavior encoded the culture's values; and they were norms and expectations other than those of the engineer's workbench. Until those values changed, perhaps with an overturn of generations, the WUI would match wilderness as the defining landscape of cultural interest.

The drivers were powerful because they came from within. The NFPA gave physical expression to that fact, too, when it constructed within its hilly headquarters a waterfall that splashes from the foyer down to the lower level where, outside massive windows, the pond resides. Flowering plants line the staggered falls, golden carp swim at the bottom. If, at 1 Batterymarch Park, a particle of the city had been brought to the country, it was equally true that a patch of nature had been brought into the built world. The NFPA had internalized the WUI. That's why the problem will not disappear soon, why the NFPA had to create a division of wildland fire, and why the wildland fire community needs its awkward liaison with the urban fire services.

THE VIEW FROM BILL PATTERSON'S STUDY

S TART A FIRE STORY in Florida, and it will eventually, sooner rather than later, lead to prescribed fire. Start one in California, and it will end, inevitably, it seems, with suppression. Start one in New England, and it will likely lead to Bill Patterson.

There is no one in the regional fire community he doesn't know, and nothing to advance the cause of fire management he has not assisted with. Hundreds of fire practitioners have trained under him. The bulk of the region's fire scientists have studied under him or have referenced his work. He collected the background data and wrote early fire plans for the major national parks. He carried driptorches to the region's archipelago of prime burning sites. He worked with the Nature Conservancy to establish a regional fire program. He carried the torch of New England fire to a national audience.

New England's spidery roads are a nightmare for someone raised out West, full of overlays, legacy routes, name changes, and destinations that often no longer exist. They are a historical artifact as much as a system for transportation, and in this they resemble the region's fire network. But no matter how many turns and missed turnoffs you take, you seem to end at Bill Patterson's study.

William A. Patterson III was born in St. Paul, Minnesota, on July 2, 1945, grew up on family stories from his grandfather, William A. Patterson Sr., about logging the Great North Woods, in this case second-growth aspen, and knew early that he wanted to be a forest ranger. He didn't become that ranger, but he went one better.[1]

The family moved often when he was young. His father, William A. Patterson Jr., had graduated after three years at the Naval Academy in 1943, became a naval officer, then, after the war, a carrier pilot until a medical discharge ended his career. Despite the moves, the family returned in summers to a hunting cabin built by William A. Sr. in northern Minnesota. The woods and what had happened to them impressed the young boy. In 1956 his father was transferred to Hingham, Massachusetts. The family arrived a year before the Great Plymouth fire burned in a rush, with 150-foot flame lengths, until it finally expired against the Atlantic shoreline. "I have this vivid memory," Bill recalls, "of being driven to scout camp through this absolutely blackened landscape." In high school he was an average student and considered the "quietest boy" in his senior class, the kind of personality that might find life in a lookout tower attractive. In fact, he built a fire lookout tower out of toothpicks.[2]

Inevitably, he went to college with the idea of being a forester. He attended Maine, and eventually came to appreciate how the frontier of logging had moved from New England to the Lake States. He notes his grandfather acquired his lands by paying the back taxes owed. That land had originally been granted to "a Dakota-Sioux halfbreed in late 1864 and eventually sold to William D. Washburn (U.S. Representative and Senator for Minnesota, founder of Pillsbury Co., first President of Soo Line RR)." The pines had been clear-cut in 1883; the slash burned, then returned mostly to aspen. His grandfather started cutting the aspen. It's now a formally designated Tree Farm. In 2005, Bill III's son, Bill IV, returned to Maine to oversee the management of more than 300,000 acres of forest land for the Nature Conservancy.[3]

The University of Maine-Orono required that Bill join an army ROTC program; but Bill followed family precedent and enlisted in the naval reserve in 1964 (eventually being commissioned an ensign upon graduating from Maine in 1967). There wasn't much fire in his program of study. The closest he came was a study of the effects of burning slash on soil properties for an honors thesis. When he returned to Minnesota for a master's, he found an active group of fire scientists. Fire ecology was a term finding its legs, and he heard its background hum, even as he pursued new disciplines—limnology and paleoecology. By now forest ecologists were discovering that fire was not so much destructive as transformative, that its removal could be as disruptive as its unwanted presence; and palynologists were discovering that all that charcoal in their cores was not simply noise, the junk DNA and lint of paleoecology, but a rich source of information about fire history. His thesis work under Henry Hansen focused on Itasca State Park, a place with a robust chronicle of fire. In 1969 he earned an MS in forestry.

The navy finally called, and from 1969 to 1972 he spent his time first aboard ship and then in Saigon working with an intelligence unit as part of the Vietnamization program. When he returned to the States, he picked up where he had left off, pursuing dissertation research at Itasca, combining paleoecology with modern disturbance ecology. In eerie echo of his grandfather, part of his work involved the killing of aspen (Agent Orange had been used since 1967 as part of herbicidal slashing, followed by burning). By now the first text on fire ecology, edited by C. E. Ahlgren and T. T. Kozlowski, was out (1974), and the hum had become a buzz. Still, what Bill knew about fire he had learned indirectly by reading *A Sand County Almanac* and Loren Eiseley and listening to lab banter and seminar speculations, even as an unhappiness with national wildland fire policy that had simmered for years was beginning to boil over. In 1968 the National Park Service recanted the 10 a.m. policy in favor of a policy of fire restoration. The U.S. Forest Service, too, stutter-stepped toward a new policy.

By the spring of 1976, with the prospect of an end to his dissertation and to work at Itasca, and with jobs scarce, he worked for the Minnesota State Planning Agency on potential impacts of a proposed copper-nickel mine; the project lasted two years. When a position opened at the University of Massachusetts-Amherst in urban forest ecology, he applied, then scrambled to complete his PhD in time to meet the hiring deadlines. Modestly—it's an instinctive trait with him—he claims he didn't know the subject, but the hirers saw enough in him to waive the details. By the time he arrived in the fall of 1978 the U.S. Forest Service had also rechartered its fire program.

None of that mattered much to New England. For the region, large fires were a memory, not an inspiration. The fire revolution didn't seem to matter much to Bill Patterson either. He spent parts of his first four summers from 1979 to 1982 in Alaska on the Seward Peninsula in what was becoming the Land Bridge National Monument, and later in the Noatak National Reserve; later, he returned for Noatak Biosphere Reserve. Fires were something that happened out West, or deep in the Alaskan bush, or back in the heyday of Lake States slashing. Fire restoration didn't seem relevant to Amherst, Massachusetts, or to someone hired to teach urban forestry.

Then he came to a fork in the road, which quickly turned into a roundabout. In the spring of 1981 Bill decided to offer a grad seminar on fire, for which he could use fields in the Quabbin Reservoir for prescribed burn training. In May 1981 he conducted his first prescribed burn. (He notes wryly that he did not evolve into prescribed fire from suppression because he has never fought a wildfire.) With fire ecologists thin on the ground, that was enough to interest the National Park Service, then mandated to write fire plans for all their holdings, even those in the Northeast.

The program started quietly with a study of Acadia National Park, the most famous site of the 1947 fires. But its scope expanded to include Cape Cod National Seashore, Fire Island National Seashore, Gateway National Recreation Area, and Saratoga National Historical Park, which wanted to burn to keep the scene roughly as it had been when the Continentals defeated the British in October 1777. In 1985, with the help of a graduate student native to Cape Cod, David Crary, he established at the Cape a complex array of plots to measure the effects of burning and mowing, both during the growing and dormant seasons, and at intervals of one to four years—all with controls. Where his ancestors had fueled wildfire with logging slash, he proposed to stoke prescribed fire with field data. Probably only the plots at Konza Prairie reap as much information. Camp Edwards on the Cape needed to quell its artillery-sparked wildfires, and sought his assistance in substituting prescribed burning. The Massachusetts Audubon Society asked him to help burn some grassy plots on Nantucket; the Nature Conservancy wanted help on Martha's Vineyard; and the program migrated to the Elizabeth Islands. He worked with Ron Myers to nurture a burn program at the Albany Pine Bush Preserve. From there he helped birth burn programs at several sites in New Hampshire and Maine (including Waterboro Barrens) and finally at the Montague Plains Wildlife Management Area in central Massachusetts. Each site fed on the others. A career switch toggled. When the National Advanced

Resource and Technology Center (later, National Advanced Fire and Resource Institute) wanted to include an eastern regional representative in its long-running course on fire and ecosystem management, it turned to Bill Patterson. He taught from 1991 to 2010.

Suddenly, he had tentacles into most of the combustible parts of New England, what might be termed the Patterson plexus; he had a topic, fire restoration, that galvanized him; he had a cause. Better, he had a teachable cause. He began to offer a one-unit undergraduate course adequate for red-carding (S130, S190) to complement the grad course in forest management. Over the years the numbers added up. With few studies predating their work, Patterson's students laid down the basic matrix for fire science in New England. He spent a sabbatical year as a Bullard Fellow, spending most of the year at Harvard Forest, with a pilgrimage to Tall Timbers Research Station, the hearth of the fire revolution. It was as though a mental magnet had strengthened, which quickly aligned all the fragments and filings pertaining to fire he had acquired over the years. He was bringing to the Northeast the field trials and basic fire science that Tall Timbers had done for the Southeast.

When asked, he points to his students. But everyone knows the fount. In 1991 the Nature Conservancy presented him with its President's Stewardship Award. The next year was the turn of the New England Wild Flower Society, with its Conservation Award. In 2010 the NPS Northeastern Region honored him with its Natural Resources Award for Research, followed by the NPS National Fuels and Ecology Award. In 2012 the Association for Fire Ecology (AFE) bestowed its highest honor, the Herbert Stoddard Lifetime Achievement Award. That was particularly gratifying because he wasn't a member of the AFE (it had been founded toward the end of his career) and because it harked back to the charismatic prophet of prescribed fire, a man whose career had led to Tall Timbers.

In October 2010 he learned that the odd twitchings and numbness he increasingly felt were the product of a brain tumor. He retired into emeritus status. The surgery was successful, though the tumor turned out to be neither benign nor malignant but anomalous, which forced him into radiation therapy. He recovered, accepted a continuing appointment, without pay, in which he has taught basic National Wildfire Coordinating Group (NWCG) fire courses and saw his final grad students to their degree. Whatever is happening he seems, somewhere, to be in the mix. His bustle remains undiminished, his enthusiasm still infectious.

In April and May 2017 he attended the Northeastern Natural History Conference in Connecticut, traveled to a workshop on pine barrens in Long Island, joined a field tour to Montague Plains for U.S. Fish and Wildlife Service regional fire management officers, then hurried to a Silviculture Institute to teach regional silviculturalists about fire use, and finally drove to the Albany Pine Bush Preserve for a North Atlantic Fire Science Exchange monitoring workshop. In spring 2017, with a former Morman Lake hotshot as a teaching assistant, he qualified 26 undergraduates as NWCG-certified firefighters, more than in any year since he came to UMass. "Seems people know I am willing to 'work' for free," he dryly notes, "and I am as busy as I can be (but not as busy as I once was)."[4]

The panorama from his study on the slopes of Brush Mountain in Northfield is a cipher for how he sees fire. He notes that the indigenous peoples called it *Mish-om-assek* ("Rattlesnake") Mountain, and the early colonists Brush Mountain. Those labels only make sense if the land was regularly burned. The closed forest canopy that blankets the slopes today support neither brush nor rattlesnakes. In the distance, by a snatch of the Connecticut River, where it bends, you can see the Montague Plain Wildlife Management Area, where colleagues and former students are trying to restore the pine bush biota.

The near view is of course actually a very deep view through history. It testifies not only to fire's longevity on the land but to active burning by aboriginal Americans. The far view is actually a vision into the future. It takes special sight to imagine in the tangle of invasives, dog-hair pitch pine, lost lupine, all amid power lines and a land often hard used, a healthy, reinstated pine bush community.

For all his commitment, Bill appreciates the niche character of New England fire. It matters, though not in the way free-ranging fires in the Bitterroot Mountains or horizon-reaching pastoral burns in the Flint Hills or chaparral fires crowding South Coast suburbs do. The last nationally significant conflagration was Maine's in 1947. The last serious wildfires in Massachusetts broke out near Plymouth in 1957, with a smaller echo in 1964. The last regional complex aligned with the drought years of the early 1960s. He makes no prophetic claims that the Dark Days are about to return, or that Maine is primed for a pyric eruption. But he knows that fire has been a persistent feature for millennia, that the last ice sheets left geomorphic patches favored by pitch pine, scrub oak, and fire-loving shrubs, and that they need fire to thrive. And he knows, as any serious naturalist in the region does, that humans have shaped those regimes. People burned, and by land use practices they have promoted fire-favored species. Strikingly, the distribution of pitch pine and scrub oak are more likely to conform to once-plowed fields than to relic patches of the Pleistocene.[5]

It can be a hard sell. New England may have once been known as the Burned-Over District, but that adjective is a past participle, not a label for the present. There are critics on all sides. Fire scientists trained in the West, which is where most fire schools reside, still want wildlands untainted by human contact. Ecologists in the Southeast accept anthropogenic burning, but anchor their philosophies with such fire-obligatory species as palmetto and longleaf pine. New England naturalists can accept landscapes as a legacy of human habitation, and they recognize the value of fire for pitch pine and blueberry, but preserved landscapes are few, fire can be unpredictable, and they imagine landscapes as akin to heritage buildings. Better to replace their inner workings than burn them to the ground and rebuild. Fire seems a specialty tool like a pruning hook, not a broad-spectrum ecological catalyst. Nobody likes smoke. Regional ecologists are more likely to unite in suspicion over fire than in its promotion. Fire might be okay in the kitchen, but not at the table amid the conversation.

═══════════

He remains, at core, a teacher. Research grants are opportunities to instruct students. Prescribed fires are opportunities to train burners. Demonstration plots are occasions to teach the public. His students are his true legacy, which makes talking to him about himself difficult because he invariably shunts the conversation to his students, who universally refer to him as "Dr. Patterson." It's an odd title—an honorific, really—for someone so congenitally soft spoken and unconcerned with the usual emblems of status. But it may be apt for someone who favors the whole over the pieces. "Personally," he says, "I'm more interested in communities than in individuals. I'm more concerned that all of human society survives than that individuals do. Because we can't any of us live forever."[6]

The community over the individual. It's a sentiment Bill Patterson believes. I'm not sure, however, that the New England fire community shares his belief with equal conviction, because there is at least one individual without whom that community might barely be said to exist.

MAINE'S EPICYCLES OF FIRE

SOME FIRE REGIONS are Copernican, with their landscapes orbiting about a fiery Sun. Fire is so powerful a presence that it shapes the space-time geometry of the ecosystems within its gravitational pull. The land is always burning (as in Florida) or wildfires routinely threaten human habitats (as in California). Others are Ptolemaic, more awkwardly centered or decentered. They are sculpted by competing rhythms, none dominant for long, and so are filled with ecological equants and cultural epicycles to account for the complicated choreography by which fire interacts with flora, fauna, and people. Their pyrogeography resists simple mapping. Their fire histories are a palimpsest, written, erased, rewritten, over and again.

Fire in Maine favors the Ptolemaic. Its landscapes bear the deep scars of the Pleistocene, have sharply etched seasons, though not organized by wetting and drying, and are constantly churned by disturbances, some regular, some idiographic. Its landscape is a micromosaic, as messy as an asteroid belt. It doesn't have vast horizons of conifers ready for conflagration. Outside of patches, reproduction doesn't demand burning. Fire is one presence among many, more like a shooting star or meteor shower than a planet, always there, but not always visible, occasionally flaring into view, full of wonder and portents. Very rarely, catastrophically, it strikes and shatters landscapes.

In Maine fire is a constant in its background ecology—always there in the biotic shadows, like a tiny pilot flame. Whether it propagates depends on how that ignition sits within an environment both oddly stable and endlessly bubbling. A boreal environment is one defined by extremes, not averages; big fires can occur within a year otherwise unremarkable for burning. The institutional environment is one that must also react to the extreme moment, to crisis and trauma, but must also reconcile the tenaciously local with the national, even the transnational. What is remarkable about Maine is how, in the end, its fire history so resembles the rest of New England's.

The rhythms compound and nest. There is a grand cycle from Pleistocene ice to Anthropocene fire. There is a colonization cycle from European contact through American settlement. Over the past century there are small cycles that swell and fall with drought and land use changes. Its fires are patchy in space. Much is only rarely combustible. Some places like barrens and blueberry fields and woods bristly with white pine are far more prone to burn than are others. And its fires are patchy in time. Some eras show a lot, and some little, in rhythms not easily coded into predictive algorithms. Pockets of pitch pine and scrub oak can burn several times a decade. The hardwood and hemlock of the deep forest burns on long rhythms of millennia. Even natural ignition is spotty, and when kindling takes, a fire must push through a broken landscape only portions of which are likely combustible. A spring fire season has fast, lighter fires; a fall one burns both deeper and broader. Fire manifests itself as its kaleidoscopic context permits.

The natural setting is a boreal biota atop the Canadian Shield; the ideal formula for fire is a lightning strike in a blowdown with a stiff wind behind it. The institutional setting has the political economy of a Canadian province within the matrix of American federalism. Fires burst forth like a plague as the European axe bit deeply into the woods. To cope, Maine gradually evolved distinctive arrangements. On its organized lands, it accepted the norms and practices of New England; on its unorganized territory, it established a special fire district; and when the motivating fire threat passed, it merged fire with generic forestry bureaus and embedded its fire task within a web of regional and international collaborators. Even its institutions are Ptolemaic.

Maine has the largest expanse of North Woods in New England, the hardest used, and the longest recorded cycle of big burns. At the time of European contact, all its mountains, save perhaps Katahdin, had forested summits. Now none do.

=====

Fire's written chronicle began with English colonization, when Maine was still part of Massachusetts. A statute for wrongful fire setting in fields and woods was enacted in 1652, and even made such arson a capital crime. But fire and axe were the primary means of converting the North Woods and barrens into habitats deemed suitable for the newcomers. Settlers girdled, felled, and burned to clear land; burned slash into potash and rough pasture; fired fallow and fields and barrens; ran flames through blueberry patches; and of course, occasionally by malice, often by indifference and carelessness, they left fire on the land that ran wild. Until major logging companies pushed into the interior, using Maine's spectacular rivers for transport, there were plenty of natural buffers and baffles. Most fires would ramble until they sank into swamps or hardwoods or were rained out. Where they threatened fields or houses, neighbors would gather to swat them out or set backfires.[1]

But as settlement grew wider, so did the fires, some big enough and horrific enough to be recorded. Before there were flames, there was smoke—the Dark Days of the early times, beginning in 1716. Then the flames arrived, beginning in 1761, followed by a cadence typical of boreal landscapes laden with logging debris: 1762, 1782, 1785. In 1763 a major fire in southern Maine stimulated settlement of eastern Maine as burned-out farmers relocated. In 1795 some 200 square miles burned south of Katahdin. In 1798 fires followed hurricane blowdowns. There were noteworthy fires in 1803, 1811, 1814 (the "whole side of Katahdin for ten or fifteen miles around"), 1819, 1822, 1823. Then came the Apocalypse. On October 7, 1825, a vast area of central Maine erupted into flame over 832,000 acres, and in New Brunswick a much broader swath, some three million acres in all. The two blowups merged in the public mind as the Miramichi fire and announced a grand cycle of conflagrations not only for Maine but for the United States. Extensive landclearing, innumerable fires set to clear away the extensive slashings, a "violent gale" that drove flame and embers before it—this was the template for more than a century of conflagrations that rolled westward with settlement. Nothing later in Maine has equaled its blast.[2]

The chronicle scrolled on. 1837: 150,000 acres along the Sebois River, started by burning haystacks. 1844: 75 square miles along the Dead River. 1858: 80 square miles along the New Hampshire border, and others around Hancock County, which reburned in 1870 and 1884. 1884: 22,300 acres, powered by logging slash and blowdowns. 1886: three fires that burned 100,000 acres along the Dead River (again). 1894: serial burns in Washington County, many spreading from blueberry barrens.[3]

By 1900 axe and fire had brought the Maine Woods to a postglacial nadir. The Pine Tree State was becoming the paper birch state or worse. It is estimated that three-quarters of the original pine had been cut, a third or a half of its spruce, half of its hemlock, a third of its cedar. Regrowth was hampered by free-ranging fires, and sometimes livestock. Still, agricultural land abandonment was underway, and old-field pine and spruce began to return. Industrial logging by rail came late: Maine's majestic river system allowed open access to the interior in ways not possible for the Adirondacks or the Alleghenies.[4]

Among those concerned with conserving something of Maine's woods for the future, the belief spread that the state would have to intervene, and that intervention should begin with fire protection. In 1891 Maine made the land agent a forest commissioner with a duty to collect statistics and, among other tasks, report on fire damages. That was five years after New York established a forest commission and the same year the national government enabled the creation of forest reserves. The trajectories of fire history in Maine, New England, and the United States were being brought into complementary orbits.

———

From the onset of European colonization, Maine's fire history had tracked preindustrial logging and landclearing. When agriculture could no longer pretend it might remake cutover lands into farms and pastures, with their associated villages, the state found itself rudely divided in half. There were organized towns and unorganized towns. The organized towns had the apparatus of rural New England fire protection, however haphazard it would seem to a later time. The unorganized towns were the domain of large landowners committed to timber, first under private ownership, then under corporate. As settlement matured, fires in the organized towns were tamed into fallow. Fires in the unorganized district were left to wind and slash.

By the time Sargent summarized Maine's fire scene in 1880, he observed that "forest fires, which formerly inflicted every year serious damage upon the forests of the state, are now of comparatively rare occurrence." Then industrialization changed the dynamics. With the move to highly capitalized pulp and paper operations and to mechanized operations in the woods, the system found itself stressed anew with few mechanisms to cope. The railways now contributed to a revived wave of slashing (in spruce) and flung sparks with abandon. The 1891 law that created the position of forest commissioner made the selectmen in organized towns and the county commissioners in the unorganized ones ex-officio fire wardens and imposed some regulations on railroads, confirming that they were liable for fires along their rights-of-way. It was a dispersed arrangement, but it did contain the rudiments of a statewide system. In 1903 the authority to appoint wardens passed from county officials to the forest commissioner. Wardens could patrol as well as respond.[5]

But big fires returned, rekindled by axe and steam. They were the same complexes that roared across the region. The Great Fires of 1903 burned 267,587 acres, over 200,000 of which gorged in the unorganized towns. ("It seemed as if heaven were burning up.") More prosaically, Forest Commissioner Edgar Ring reported that the season would "go down in history as one that has never been equaled and it is hoped never will be repeated." Then it was repeated as another round of Great Fires returned in 1908, this time for 142,130 acres, of which nearly 100,000 was in the unorganized townships. The embryonic institutions of 1891 collapsed. The organized towns barely coped, and often failed. The unorganized towns could do little but stand aside from the flames. After Somerset County suffered 47,120 burned acres, lawsuits sought damages from the Canadian Pacific Railway as the source of ignition.[6]

After the 1903 outbreak, some minor adjustments were made. Two years later the first of what eventually became an elaborate network of fire lookouts ("watchmen") was erected at Squaw Mountain. But it took the second outbreak, smaller in area but larger in shock, to stimulate reform. The upshot were laws that clarified issues in the organized towns (extending to cities as well), that established a Maine Forestry District (MFD) that brought system to the unorganized towns, and allowed the governor to close hunting seasons (and thus campfires) during drought. These mostly northern lands amounted to a little over 10 million acres or half the state.

The Maine Forestry District involved a collaboration between the state and the landowners financed by a property tax. Here was Maine's version of the timber protective associations and the state-county fire arrangements that appeared in the American West at the same time. What Maine contributed was a stronger sense of state involvement. It helped that landownership resided in relatively few hands, that Maine's timber economy was corporatist (closer in style to Canada's than to most of the United States), and that owners accepted the need to contain wild-fire. The response, unsurprisingly, was minimal—it was hard to justify an annual program that could meet the worst fires. Most years found the program overstaffed; the bad years, woefully inadequate. Still, the reform brought system to lands otherwise beyond the reach of organized fire protection.

The lands of the MFD were a historical curiosity of Maine's land tenure history. Originally, when it was the Eastern District of the Commonwealth of Massachusetts, some 19 percent of its landed estate was sold or granted to the private sector. But after the Revolutionary War, such were Massachusetts's debts that it had to sell its principle wealth, land, to pay them. An aggressive program of disposition began. Maine was split off from Massachusetts in 1819, then admitted as a state in 1820. Maine received less than a third of the lands under its jurisdiction; Massachusetts had sold or granted the rest, and continued to hold some through 1853, by which time Maine had purchased the Massachusetts lands and resold them through public auction. The oddity of its split from Massachusetts, as with West Virginia later, left no lands with the federal government. The sales went on, and on, not as in western homesteading to small families but mostly to wealthy individuals who acquired large blocks in the great timber belts. Some 700,000 acres were granted to the European North American Railway Company. By 1878 the frenzy ended. All but 400,000 acres of public lots in unorganized towns remained, along with a few coastal islands, although even these lands had their timber rights sold.

This left Maine in peculiar circumstances. Most western states (and a few others, like Florida) still held large blocks of unpatented public land under federal jurisdiction. Much of these would be organized around a national program of state-sponsored conservation. (The year Maine sold off the last of its lands marked the publication of the *Report on the Lands of the Arid Region of the United States*, a marker in western land reform.) The national government, primarily through the U.S. Forest Service, brought fire protection. Texas had never surrendered its public lands to the national government upon admission, and so found itself having to manage its western lands on its own, and struggled when fires returned in the 21st century. States without a federal land presence were likely to be taken over politically by commodity producers—Texas by cattle and cotton, then oil and gas; West Virginia by coal; and Maine by timber. Control over major resources like minerals, water, and wood passed into private hands.

This meant Maine would have to devise a fire protection system on its own. New York had retained lands in the Adirondacks, and would purchase others, and arranged a forest commission to oversee fire control. Many of the Lake States followed its example, though with assistance from a federal presence. Maine had a vast estate that logging was making increasingly prone to explosive fires without a public institution to administer it. There was a reason the lands were called "unorganized," and their scale—some accessible primarily by canoe—made the projection of state institutions difficult. Maine's solution was corporatist, to work with the powers in place, not unlike what happened in Texas and West Virginia, or what evolved in eastern Canada.

In normal years the system worked fine. In big fire years it crashed. Nor was it obvious who would be responsible for installing the shared infrastructure of roads, lookout towers, telephone lines, power pumps, firefighting tools, prevention campaigns, and so on. Landowners accepted the fire tax on property where they saw a threat; they were reluctant to invest in public goods. Maine needed something else to boost the system.

That came in 1911 when Congress passed the Weeks Act. The law spelled out an institution for federal-state cooperation through grants in aid to promote state forestry bureaus, especially around fire protection; and if the state could not rally the will or resources, it allowed for federal purchase of lands for national forests. Maine quickly joined. The money did exactly what the program intended: it jump-started the state's investment in fire protection and gave some stiffening by appeal to national standards. It joined Maine to something more than its own inbred history: it made it part of a national program for conservation. It did not lead to federal lands; that enthusiasm went into the Green and White Mountains in Vermont and New Hampshire. It did leave the State of Maine to broker between national concerns and local landowners.

The fires continued. There were a handful of big years; ironically, 1911 was the worst year of the rest of the century, save for 1934 and 1947. There were years swarming with fires that did not blow up. Forest Commissioner S. T. Dana reported that in 1921 not a day passed from April 28 to September 24 that did not record a fire. The threat was constant, like a chronic plague reservoir, only waiting for the right but rare circumstances to allow it to burst forth.[7]

But federal assistance grew, too. The 1924 Clarke-McNary Act replaced the Weeks Act and injected a major source of additional aid. Then the CCC arrived. What had been a slowly evolving infrastructure sprang up in a handful of years. The CCC erected 107 foot bridges, 240 vehicle bridges, six lookout towers, 482 miles of telephone line, 838 miles of trail (for foot, horse, and truck), reduced hazards along 551 miles of road, and spent nearly 30,000 man-days fighting fire. Conceived as a reserve force ready for backup, too often the CCC became first responders, saving the MFD and the state precious fire-suppression dollars. The contribution endured. Austin Wilkins, long-serving forest commissioner, observed that well after the CCC had concluded, "it was surprising to learn of the number of people who responded to the call for forest fire fighters whose experience, learned while members of the CCC qualified them for certain positions in the fire suppression organization."[8]

Then came the firestorm of 1947. Its center lay in the southwest, overlaying nicely with the first big fires of Maine history from 1761 and 1762. A series of fellings and blowdowns supplied the fuel: from the 1938 hurricane, from a 1945 snowstorm, from accelerated logging during the war and then afterwards. This time the organized towns bore the brunt. A few fires, having skunked around for days, blew up on October 23 when gale force winds swept over the land. The damage was great: 15 dead, 220,000 acres burned, 2,500 structures lost, nine towns obliterated, four towns damaged, and half of Acadia National Park incinerated, along with much of Bar Harbor and all of the Jackson Cancer Laboratory. The shock was greater. Almost 40 years after 1908, after the creation of the MFD, after the Weeks and Clarke-McNary Acts, after the CCC, after decades of steady improvement in which wildfire seemed closer to the Pleistocene than to post-WWII America, conflagration had returned. "Never in the history of the state of Maine," muttered Austin Wilkins, director the Maine Forestry District, "have a series of forest fires caused such devastation and privation." In relative terms the damages were likely greater than in 1825.[9]

Obviously more was needed, beyond what Maine or the other states affected by the outbreak could provide on their own. The region reached back to a provision of the Weeks Act that allowed for interstate agreements for fire protection—a recognition that eastern states in particular did not have a federal land base or fire-hardened bureaus from which they could draw help in an emergency. What emerged was the 1949 Northeast Forest Fire Protection Compact. New states entered, so did Canadian provinces, along with several federal agencies. Collectively, they created a regional reserve apart from the national cache.

Already Maine had entered into the cycle of declining fire that characterized the region. After 1947, however, the trend steepened. Like a boa constrictor squeezing its victim with each exhalation, with each pause the system took more fire away. For over a century the pressures of settlement had wound the environmental mainspring for fire tighter. Now, the hand lifted; the fire's clockwork ticked more feebly and slackly. The number of fires jumped sharply after 1947, probably as a result of improved detection, but burned area plummeted. The decade from 1901 to 1910 burned an average of 61,127 acres a year; the decade from 2001 to 2011, 908.[10]

In 1972 legislation dissolved the MFD and absorbed its duties into a Bureau of Forestry, later the Maine Forest Service (MFS), more obsessed with spruce budworm than with flame. Fire was one disturbance, one task, among many, not a challenge around which an agency might gather, precipitating like crystals on a string. A fire lookout network that Maine had proudly declared the first in the nation began to come down. Maine had taken a distinctive turn in its fire history, but a century after 1908 its fire scene looked much like the rest of New England's, for many of the same reasons.

What would it look like a century after 1947? It will probably look like the rest of New England. The deep drivers are demography and economics. Maine is depopulating, and what remains is aging in rural townships or moving to cities. In 2017 Maine had roughly twice the population of Boston, and the same population density as Colorado. At 43.5 years Maine has the highest median age of any American state. Some towns can no longer field a fire department; some townships have petitioned to move from legal status as an organized township to unorganized territory under state control. Some timber companies are divesting, removing roads and bridges, the infrastructure that had made the land accessible, while putting those holdings on the market for purchase by either developers or protectionists. Slowly, inexorably, much of Maine is rewilding. The vision of a restored Great Northern Forest is a Northeastern echo of the Great Plains' buffalo commons. What happened in 19th-century New England with land abandonment is happening again in early 21st-century Maine.

One consequence is a growth of feral land. Even where former timberland converts to subdivisions, the houses are typically dispersed, allowing woody fallow to flourish. A good fraction is going to public land, or private land that serves a public purpose. At one point virtually all of Maine was privately held. Since then public lands have grown. Part of White Mountain National Forest spills into Maine, an early contribution of the Weeks Act. Wealthy donors gave it Acadia National Park and Baxter State Park, and more recently the land for Katahdin Woods and Waters National Monument. The Fish and Wildlife Service has half a dozen refuges. The State of Maine and municipalities own preserves. And NGOs have become a major player. The Nature Conservancy manages 75 reserves with 300,000 acres, and assists others on a total 1.7 million acres. With more land being sold, the prospects for increasing acquisitions for nature preservation are rising. All in all perhaps 7 percent of Maine is now formally public, though easements and NGO-controlled lands push that number to 21 percent. Conservation easements and working landscapes seem to sit easier with Mainers who worry more about public access than public ownership, and who seem more inclined to work the land than leave it alone, even if the projects advance ecological goods and services rather than commodities.

But even as public land was swelling, support for public fire services was shrinking. The Maine Forest Service had half the staffing of 20 years earlier, volunteer fire departments were dying off due to aging populations and a decline in volunteers, land previously in organized townships was being transferred to the unorganized territories that became the responsibility of the state. Better training, more sophisticated technology, keen prevention programs—the Maine Forest Service has been able to hold ground, but not advance. At some point smaller budgets and larger missions will cause a switch to flip. The threat is not so much one monster fire as a swarm of smaller ones, not all of which can be handled at once and some of which will grow big. The MFS can appeal to collaborators, but when conditions favor big burns in Maine, they also affect New Hampshire and Quebec, and it is initial, not extended, attack that matters most. The agency's remarkable success in excluding fire has paradoxically convinced the public that Maine has no fire problem, which is like arguing that the reduction of measles means vaccination and public health services are no longer needed. Changes in land use change the opportunities for fire. Over the past 400 years fires have largely fed on the carrion left by the axe. Now fallow land offers fuel for feral fires. The new estate promises to more approximate what existed before—not restore that former landscape (too much has happened), but create conditions that might allow lightning and stray anthropogenic sparks to reassert a presence and even challenge fire suppression as the sole strategy for fire's management. The issue took dramatic form when lightning started a fire in Baxter State Park on Sunday, July 17, 1977.[11]

The circumstances help explain both the rarity and the tenacity of lightning fire as an ecological presence. The fuels dated to Thanksgiving Day, 1974, when heavy snow and strong winds caused a blowdown. Some of that windthrow—nature's slashing—lay within lands belonging to Great Northern Paper Company, which promptly sent in heavy machinery to clean it up. The rest lay in Baxter State Park, an extraordinary bequest from a former governor, Percival Baxter, who in 1931 had gifted 6,000 acres (what would eventually become 200,000 acres by the early 1960s) around Mount Katahdin to the people of Maine to be held in trust by the state and maintained in a "Natural Wild State." It was a gesture of penance for the havoc caused by concentrated private capital. (Much of the early acquisitions lay in the burn scar of the 1903 fires, which may be why timber companies were willing to part with it cheaply.) Mount Katahdin would do for Maine what the Adirondacks did for New York. The Baxter State Park Authority was permitted to "to clean, protect and restore areas of forest growth damaged by ACTS OF NATURE," fire among them, while maintaining the park's purportedly pristine character.[12]

The idea of wilderness was a work in progress. When Baxter made his first purchase in 1930, the Wilderness Society was five years away from its founding charter. When he made his final purchase in 1962, the Wilderness Act was two years in the future. What managing wilderness meant was undetermined, but most partisans agreed that it meant a light hand, no roads, and no heavy machinery. It meant not doing in the park what was done around it. Great Northern salvage-logged. The park moved more cautiously, alarmed about the fire hazard, both to the park and to neighboring lands, yet struggling to find a company willing to clean up at a cost the park was willing to pay. That set up a lawsuit over the right to intervene at all. The final court ruling allowed the park authority to proceed but without heavy equipment. That decision came nine months after lightning had sparked a fire in the blowdown.[13]

There were two fires, one extinguished, the other so encumbered by debris that rangers dared not risk sending crews in that night, instead evacuating the Abol campground, and preparing for a dawn attack. Wind and fuel pushed the fire briskly, while blowdown and rocky terrain prevented bulldozers from working effectively. Flames jumped early control lines. By midday Monday there were 100 firefighters, four Beaver and one CL-215 air tankers, one helicopter, eight bulldozers, three skidders, four water tenders, and a dozen portable pumps. The numbers went up as the burned acreage relentlessly rose. By the next day the fire had grown to over 1,000 acres, and added another thousand the day following. A cold front blustered through, further fanning the flames and hurling spots. The Beavers proved ineffective; the CL-215 was recalled to Canada to address fires there; the dozers broke down on rocks or sank in mud. At one point crews feared "for their safety." Burnouts were dismissed as ineffective. Fire weather forecasts proved inaccurate. Then the fire calmed, control lines were completed, and a week after ignition, crews began to demob. The final size was estimated as just under 3,500 acres—a thousandth the size of the Miramichi complex and a twentieth that of the 1903 burn, but a monster by modern Maine standards, and the largest since 1947.[14]

As challenging was the public firefight that ensued. There was little doubt that a serious salvage and cleanup operation could have reduced the fire hazard—Maine had interacted with its woods by appealing to the axe since Europeans had first arrived. The concern by critics interested in a different relationship was whether such a project would do more damage than a fire and might set precedents for future interventions. (In fact, the dozer lines are still visible from the air 40 years later.) There was less discussion about the ecological value of the burn since fire was not considered essential to the park's ecology. What mattered was that the response to a fire not violate the park's wilderness ethos.[15]

The meaning of wilderness resided in Baxter's deed, not in national legislation or the declarations of wilderness prophets. The fire occurred only three years after Teton National Park had overseen a 3,500-acre Waterfalls Canyon fire that burned amid considerable media attention, and two years before the Ouzel fire blew up in Rocky Mountain National Park and threatened surrounding communities. Only the year following would the U.S. Forest Service formally adopt fire's restoration as an end, and fire by prescription as a means. The mesmerizing spectacle of free-running fire in Yellowstone was 11 years out.

So this was a very Maine wilderness fire fought with Maine methods and debated in terms that resonated with Maine's history. What it illustrates for Maine's fire history is the way new land uses could combine with old ignition sources to challenge fire agencies and thinking about the place of fire. If fire compelled new thinking about the park, the park also forced a rethink of inherited notions about fire. This was not the same kind of fire that had threatened Maine's lands in 1903 or 1908 or even 1947.

Traditional wisdom holds that lightning fire is negligible, about nine fires a year, though surely among those large numbers of fires of "unknown" origins there are a few more. Mostly those fires had little room to roam: they soon ran into rivers, lakes, northern hardwoods, clearings, or other barriers. In 1952 lightning started 209 of 301 fires, and because they were distant from settlements, they were the most damaging in 50 years and the most costly to fight. Midcentury records (1936–1972) show an average of 70 fires per year. Whether or not climate change will boost those numbers is unclear, though recent years have seen drier summers and more lightning ignitions. What matters is that lightning will find more areas amenable to burn. What matters is that it is always there and has certainly started fires since the ice left. The Baxter fire will surely find relatives.[16]

The other issue is the active reintroduction of burning for ecological purposes. If lightning fire might reinstate one side of the precontact regime, anthropogenic burning to promote habitat reinstates the other. Pre-Columbian peoples burned in barrens and pitch pine-scrub oak and blueberries and sedges, and left fires along trails and piney valleys, and to maintain those biomes modern managers will have to restore fire to fire-hungry sites. Southwestern Maine has some scattered swathes of grasslands and a few relatively large patches of barrens and pitch pine-scrub oak forest, part of that extensive archipelago that stretches along the northern Atlantic Coast of North America.

The Nature Conservancy's Waterboro Barrens in southwest Maine is notably large (2,140 acres), and widely regarded as a premier site. Unlike most barrens, it sprawls over moraine rather than outwash and has few mesic invasives. Its history is well documented, it has not been plowed, and it has historically burned. It's the site of one the earliest reported big fires in Maine (1762) and the last (1947). For the next 60 years fire had—mostly—been kept out. Scrub oak thickened, open canopy closed, and the habitat degraded.[17]

The conservancy acquired the land from a failed subdivision in 1993–94. It spent 10 years poking, prodding, studying, trying to learn how to put fire back in. Restoration relied on the usual formulas. Thin the overly dense pitch pine and open the canopy (that prescribed clearing can help pay for other treatments). Mow fuelbreaks and scrub oak. Target old blueberry fields. Burn. Begin in the savannas and blueberries, perhaps on a four-year cycle. Move into the forests as possible. Start with easy fires in the dormant season. Push the prescription toward hotter burns as experience and opportunity allow. Widen the range of applications. Expand the effective range of the preserve by connecting with state habitat lands and conservation easements nearby. Burn some more. Work with collaborators—lots of partners. It's long-settled country, but fire has faded from rural consciousness, and likely never entered the minds of exurban newcomers except as viewed on screens as the latest disaster from California. Apart from its blueberry barrens, Maine has scant history of a thriving fire culture; and as blueberry production declines, even that vestige is fading. Cultivate an alternative. Involve local fire departments. Establish solid examples of good fire. Find ways to reconcile fuel reduction with ecological enhancement. Find means to enlarge the prescriptions for burning and earn the respect of local fire agencies. Burn prudently. But burn.

In brief Maine can expect more ignition, more lands amenable to it, and more reasons to reintroduce fire. Lightning is likely to increase, technology keeps inventing new ways for fire to trail people's passage (from catalytic converters to power lines); and prescribed burning will replace rural fire habits. Those sparks will find new sites to kindle not only in public lands and protected preserves, but in that woody fallow that flourishes as more timber holdings yield to rewilding and rural land to suburban sprawl, and as climate change modulates the rhythms of wetting and drying. The capacity to hammer every fire quickly, which has been a hallmark of the MFS, will soften. Some fires will escape. Here and there, like an outbreak of a plague thought buried in the past, larger fires will flare up. Fire control will have to rethink its tactics. The public will have to reconsider its expectations. Mostly the shifts will be incremental. Maine won't rival Florida for deliberate burning, California for explosive conflagrations, or the Flint Hills for a rooted fire culture. But in some form of other, flame seems destined to return, and fire agencies will surely seek to nudge themselves to prepare for that future.

The recovery of fire even extends, if circuitously, to literature. In 2017 Annie Proulx retold the 1825 fire in *Barkskins*, her multigeneration novel about logging, and Anita Shreve rekindled the 1947 fires in *The Stars Are Fire*. Those are not the venues or return intervals or themes most fire managers would wish for. They leave fire as the flames of perdition, in the past, amid slash, whether natural or social, as a disaster visited upon the sins of inhabitants. But it is part of a more general restoration, a reinstatement that puts fire back into cultural memory. If fire is going to return to Maine's landscapes, it will do so in the hands of people who appreciate its value not only as an ecological process but as a social presence that makes Maine what it is.

———————————

Maine's early fire crisis was a wave train of bad burns gorging on the abandoned offal of landclearing and logging. Fire was one of a chorus of hazards that threatened how Mainers lived on the land. As Maine's economy shifts, or tries to shift, from commodities to services, it is again changing people's relationship to land, rivers, and creatures. The restoration of wildfire will join new invasives, climate-kindled hazards, and the pathologies of urban sprawl; the restoration of ecological fire will align with decommissioned dams, rehabilitated salmon streams, and renewed soils. Maine's people will adapt to new circumstances. So will fire, either by default or by design.

Now the equants and epicycles that map fire in Maine are shuffling into new alignments. Maine can expect fires, and in some places needs them. Certainly its fire agencies like those throughout the region do, if only to maintain their credentials. They can get some qualifications by doing more prescribed burning. Mostly they have to go elsewhere to keep their experience current—the region first, then the country when snow is on the ground in Maine and the Southeast wants to prescribe burn and when big callouts boom out of the West. In 1891 Maine had no rural fire institutions, save a state fire organization in name only. Today it has several, all securely positioned in a web of state, regional, national, and international fire relationships. Then there were few fires it wanted in the woods, and it devised means to suppress them. Now it is discovering it wants something else out of its woods and maybe its fires.

In the early 20th century Maine's economy rested mostly on timber; it promoted good fire for burning blueberries and recycling fallow and suffered bad fire as a plague of conflagrations burned through slash. It created agencies, notably the Maine Forest Service, in response. In the early 21st century its economy is pivoting toward services, cities, and amenities. These are profound changes: they will likely come slowly and by twitches rather than abruptly by torquing, but they will alter the landscape. Legacy institutions may find that they have to edit their charters to match the fires coming at them, while newly emerging institutions may discover they need to find fires to match their missions.

COLLECTIVE SECURITY

The Northeast Forest Fire Protection Compact

I N 1947 MAINE BURNED. From October 13 to October 27, 200 fires, mostly along the coastal plain, burned across 220,000 acres, nine towns, 851 homes and 397 seasonal cottages, and half of Acadia National Park, slashed through Millionaire's Row at Bar Harbor, and consumed the Jackson cancer research center. The fires left 2,500 people homeless on the cusp of winter. This was, fundamentally, a rural fire complex kindled by traditional rural causes that overran towns and hamlets, not cities. It was fed by fuels drained by 108 rainless days, stuffed with spillover from the still-uncleared debris of the 1938 hurricane, excess felling during World War II and afterward, all overlaid by kill from a severe 1945 snowstorm. Scores of small fires tied up community departments and volunteer response, and then the Great Wind that accompanied a cold front on October 23 blew nuisance fires into a mutant conflagration. If the 1825 Maine fires began America's long cycle of settlement burning, the 1947 Maine fires might be said to conclude it.

In 1825 there was no organized response to the flames, no mechanism to care for those afflicted, no institutions to oversee rebuilding and prevent a reccurrence. The logging that fueled the burns moved on. Once burned, the land was open to farmers, at least until they failed, and to a returned forest. The fires became an artifact of an ancient past, like a faintly remembered plague. In the century that followed, fires continued of course, but they resembled seasonal flu, some years worse than others, but none of them catastrophic. Locals handled the outbreaks with the equivalent of Auntie Alice's home remedies.

In 1947 many small towns, though not all, had volunteer brigades; ruralites turned out to help neighbors, and volunteers appeared from Massachusetts and New Hampshire; the Maine Forest Service could offer some assistance; the federal government sent the light cruiser USS *Little Rock* to Portland so its crew could assist and later allotted two B-17s to drop dry ice over York County in an attempt to stimulate rain. Afterwards the Red Cross mobilized, Governor Hildreth appointed a Fire Emergency and Information Committee, the Maine Extension Service proffered advice and aid, donations poured in, and the federal government offered assistance from the War Assets Administration to the Veterans Administration to the Reconstruction Finance Corporation. Congress appropriated monies for Acadia National Park to clean up. In 1947 Maine was far better positioned to cope with a blowup than it had been in 1825 (barely five years after statehood). Yet the assets at risk were far higher, and proportionately the state had again been nearly as helpless. A modern world required greater investments in fire protection. Postwar America, bent on replacing the rural with the suburban, demanded still more. Postmortems on the fires flourished, all of which recognized the extraordinary circumstances that powered the conflagration, none of which thought Maine's response adequate. The blowup was unprecedented, though not unpredictable. In a summary report Deputy Forest Commissioner Austin Wilkins wrote, "It is frankly admitted that there was no fire action plan to meet such a disaster." It was not so much lack of resources but their randomness and lack of coordination that crippled efforts.[1]

That was the conclusion: the failure was systemic. It was not just that crews, engines, pumps, and hand tools were insufficient (which they generally were). It was that what did exist, and what might be called up, were disorganized. They were no means to coordinate assistance from outside local townships. There was no ability to muster fire control forces where they mattered most. There were no plans based on predicted fire behavior, or even weather reports. There was no controlling authority. Until October 1947 there had not seemed any reason for one. Big burns came as often as hurricanes. But what happened once could happen again, and there were actions to prevent and suppress fires that were not available for hurricanes, which happened only once a generation (or two). Maine suffered far worse than adjacent states with similar environmental conditions. Mainers demanded a political response appropriate to the disaster.

=====

It took two years, but they got one.

Reform began within the state. On January 13, 1948, a statewide conference convened in Augusta to review and recommend corrective action, among them the creation of volunteer fire departments since most of the burning had occurred in the organized townships. In 1949 the legislature enacted a rash of bills that made the Maine Forest Service the focal organization. It moved beyond the unorganized territory that had, for 40 years, been its legislated habitat into the organized townships. It established a chain of control from local wardens. It created a state master plan for fire control, helped to funnel war surplus equipment from the U.S. Forest Service to local brigades, sparked reform to regulate slash burning—worked, in brief, to create a modern fire protection system. It could not cope with conflagrations, but it could hope to contain the number of laissez-faire burns on bad days and so dampen the likelihood of future blowups.

But the state could not alone cope with black-swan emergencies that might come outside the memory of two or three generations of Mainers. It needed to reach beyond its borders. One obvious channel was the U.S. Forest Service. In its own report on the calamity, the service, "as a cooperator with the state," accepted that it too "must fairly accept a part of the blame for not having foreseen the danger of such a calamity." Still, without lands of its own, the USFS could have only a limited, abstract presence, a liaison to channel federal assistance through the MFS. In fact, the federal government had little land of any kind in Maine, which made obvious the need to rally neighbors. As early as November 1947, with some peat burns still smoking, the governors and state foresters of the New England states met to discuss what forms regional cooperation might take. They all recognized that none of them had the resources to cope with an outbreak on this scale and that none had federal land holdings that could serve as anchor points. They would have to help one another.[2]

The upshot was a proposed interstate compact, an idea the Council of State Governments endorsed. Such an arrangement could require

Congress to enact approving legislation, which it did on June 25, 1949. The original group included the New England states and New York (Rhode Island had to wait a year for its legislature to meet). By now rural fire protection was being enveloped by a national interest in civil defense, particularly measured against predictions that a future war would be a fire war. On August 29, 1949, the Soviet Union had exploded its first atomic bomb. On June 25, 1950, a year after the Northeastern Forest Fire Protection Compact was authorized, the North Korean army invaded South Korea. The next month the NFFPC held its first meeting. In August 1952 it mobilized to assist Maine when drought again returned problem fires.

Even regional resources might not be enough. Congress had authorized the NFFPC two months after the founding of NATO, and so, like the United States, the northeastern states looked beyond national borders. In 1952 Congress authorized the compact to expand into Canada. Canadians were interested but wary, however, and Quebec did not join until 1969, and New Brunswick in 1970. Another surge followed as Nova Scotia signed on in 1996, Newfoundland-Labrador in 2007, and Prince Edward Island in 2017. Meanwhile the concept of membership has expanded to include nonvoting associate status. The U.S. Forest Service, through its Green Mountain and White Mountain National Forests and Branch of State and Private Forestry, enrolled in 2007, and the National Park Service and U.S. Fish and Wildlife Service in 2011. The Fire Department of New York City joined in 2015.

But there was more still. The compact idea itself began to multiply. Other states with limited federal lands soon emulated the NFFPC, giving rise to Southeastern Forest Fire and South Central Compacts. By 2016 there were eight compacts. Only six American states did not belong to one, while three states (Virginia, West Virginia, and Mississippi) and one province (Saskatchewan) belonged to two. The Great Lakes and Northwest compacts also crossed into Canada such that only the Arctic territory of Nunavut lay outside an agreement. In 2012 the association began to organize among themselves into an Alliance of Forest Fire Compacts. A parallel, shadow infrastructure for fire protection apart from federal or national had emerged. The old hierarchies, a triumph of national integration, blurred into tangled webs.

The science of fire ecology does not include the study of institutions, but it should. The effects of the 1947 fires have rippled out from York

County to affect how fire appears in Florida, Oregon, Alaska, and from Prince Edward Island to Yukon Territory. Those institutions have practical consequences for how fire appears (or doesn't) on a lot of landscapes.

—————

What does it mean?

The significance of the NFFPC begins of course in the region, and it means more than access to materiel and personnel. It does that, of course; and requests began as early as 1952 when New Hampshire and Massachusetts sent aid to a Maine again blighted by drought. The compact helped all its members through the tough drought of the early 1960s. It addresses that most basic of limitations for fire management: lack of capacity and its dark twin, lack of experience. Through a slate of working groups, the NFFPC keeps members current in prevention techniques, technologies for fire suppression, operation protocols, and fire science—for which it plans to co-convene with the North Atlantic Fire Science Exchange. It substitutes institutional intensity for declining fire threats.

The Northeast recapitulates a fundamental paradox that can suffer from its success. In naturally fire-prone landscapes fire protection allows fuels to build up (and ecosystems to decay). In landscapes not prone to routine fire, fire protection destroys the argument for its existence. The problem has gone away. The more the compact succeeds, the fewer big fires it must deal with, and the fewer its bad fires the poorer its political profile. Wildfire appears to be an old problem solved to a public ready to move on to more pressing issues. The tendency is strong to pull back, to rely on the group as a whole. The analogue in public health would be to reduce vaccination programs, to depend on herd immunity. Haven't measles and whooping cough gone away?

In 2011 the compact contracted for a stress-test study. Over the past century big burns had been scrubbed from the landscape. Nothing equivalent to the 1947 blowup had followed. (The 1957 Plymouth fire and 1995 Long Island fires were two orders of magnitude smaller.) Plenty of fires started, but they were quickly squelched. The system was larger than the challenge. Responders could match ignitions. The occasions—dry spells with high winds—that might leverage multiple starts in a conflagration were few, and they were local in area and short in duration. The lands outside formal protection shrank. A blowup might occur in days, or per-haps a week, not over months. Big fires were high-impact, low-risk events. The wildland fire community has never solved the big-fire conundrum; but where fires are abundant, the resources are at hand to cope, while where fires are little more than nuisances, the resources tend to be sparse and scattered. The stress-test exercise was inconclusive, except to note that the potential for rare eruptive fires was still latent and that the values at risk were far higher than in the past. The real stress was less the ability to respond to a callout than to maintain capacity and continuity.[3]

What the Northeast has are institutions. Bringing first-order fire pro-tection to places that had none dramatically reduced the potential for bad burns, and then the states (and provinces) created a second-order system for collective security. The region substituted agencies and relationships for mechanical muscle and staffing numbers. Like simple public health measures for ensuring inoculation and basic sanitation, the effects could be dramatic. But then institutional innovation has long been something of a northeastern specialty.

Still, the compact's experience flies in the face of business theory and government trends over the past 40 years in which self-conscious disrup-tors have sought to remove layers from institutions, to flatten bureaucra-cies, to declutter procedures, to disintermediate organizations, to promote the smaller, the nimbler, the more responsive. Popular theory favors the startup over the establishment. So, too, the NFFPC looks to mergers and acquisitions, even at the cost of cultural clashes that will require plenty of care and feeding to smooth over. What does the FDNY have to do with state forestry in Vermont? Why should Massachusetts care about fire in Newfoundland? The NFFPC and the continent-spanning association of forest fire compacts would seem to argue for more institutional density at a time when theory and politics seem to want less. They argue for institutions amid an anti-institutional age.

That observation may miss the point, however. The NFFPC is not mindlessly adding middle managers into a new, layered hierarchy: it more resembles a web. The ties are personal as much as prescribed by organiza-tion charts. The cost of wrangling so many parties with so many interests can be daunting, but the future of American fire has been demanding just such relationships over the past few decades, and will likely demand many more. Those interconnections, not just engines and airtankers, are what matter most. New England's town-centric management of woods can seem baffling and archaic to outsiders, who may dismiss the need for complex connections. Yet even the public lands of the West are thick with stakeholders since they belong to everyone. The Four Forest Restoration Initiative in northern Arizona may involve only national forests, and so one administrative agency, the U.S. Forest Service, but some 20 parties have claimed a place at the table. The resulting negotiations will determine the success or failure of the project. The National Cohesive Strategy is attempting to bring together volunteer fire departments, federal agencies, state bureaus, research projects—this looks a lot like the Northeast Forest Fire Protection Compact. The NFFPC shows how such relationships can work. They are the future of landscape fire in America.

The states and provinces have plenty of practical reasons to participate in NFFPC. New York can gain quick access to single-engine air tankers and CL-415s out of Quebec. New Hampshire can learn from Massachusetts about converting military surplus trucks into water tankers. Connecticut can exchange its experience converting shovels into fireline rooters with New Brunswick's clever appliance for opening fuel cans. During the 9/11 disaster the FDNY learned the value of the incident command system, developed for wildland fire, and wants opportunities for its incident management teams to stay current, and wildland fires offer more opportunities than Gotham does. Each member contributes according to its abilities, each gains according to its needs. It's not about hurling hotshot crews into remote mountains. It's about getting along with very different others, some large, some small, some near, some far, toward a common cause.

That's a lesson the Northeast had to learn a long time ago. You don't need fleets of Type 6 engines and phalanxes of Type I crews to contribute to the national system. Sometimes—maybe most of the time—before you can do something effective in the woods you need to work with people. Fire management is about getting the right fire on the land, and if your land is complex, your tenure and governance systems fine-grained and layered, your capacity a function of cooperation, you need to talk and listen before you act, and listen more than you talk. That's a skill the Northeast has learned. The rest of the country would do well to listen in.

WESTWARD, THE COURSE
OF EMPIRE

Westward the course of Empire takes its way.
The four first acts already past,
A fifth shall close the drama with the day:
Time's noblest offspring is the last.
  —BISHOP GEORGE BERKELEY (1685–1753), ON
  BRITAIN'S NORTH AMERICAN COLONIES

I N SEPTEMBER 1919, still recovering from his war wounds, Ernest Hemingway went to the upper peninsula of Michigan to fish. He later wrote a two-part short story, "The Big Two-Hearted River," based on that experience, published in 1925. It opens with the panorama of a burned-over landscape.

> The train went up the track out of sight, around one of the hills of burnt timber. Nick sat down on the bundle of canvas and bedding the baggage man had pitched out of the door of the baggage car. There was no town, nothing but the rails and the burned-over country. The thirteen saloons that had lined the one street of Seney had not left a trace. The foundations of the Mansion House hotel stuck up above the ground. The stone was chipped and split by the fire. It was all that was left of the town of Seney. Even the surface had been burned off the ground.[1]

The fire, he realizes later, must have come "the year before." He scans the countryside, "burned over and changed," such that even the grasshoppers are sooty, but decides "it did not matter. It could not all be burned." He seeks out unburned patches, a refugia for the soul. He finds a clean stream with good trout.[2]

The burned-over woods are a metaphor, an external manifestation, of the internal state of the protagonist, Nick Adams, burned out by the war. His fishing expedition is an attempt at recovery. Revealingly, he leaves the river when it splits and enters the dark quiet waters of a swamp. He might try the dark place later, after he had regained competency and sanity. "There were plenty of days coming when he could fish the swamp."[3]

But the burned-over landscape was not just what literary critics call an objective correlative. It was an objective reality. For decades the region had been ruthlessly cut and fecklessly fired. The reckoned fire year, 1918, had seen massive burns, some of which had incinerated Moose Lake and Cloquet, Minnesota. And those fires were the final tremors of an earthquake of axe and torch that had transformed the landscapes throughout the Lake States, even seemingly remote settings such as Seney. They had slicked off cutovers and scoured out cold swamps. Now the landscape, like Hemingway's traumatized hero, was struggling to recover. In fact, a few patches had survived. Rehabilitation would take a lifetime. The scars were deep. Even by the time Ernest Hemingway killed himself in 1961, recovery was slow. The vanished woods had banished fire. The wreckage of axe and torch remain today.

The Lake States were where the Northeast went to slash and burn after it had felled and fired its own woods. Westward, as Bishop Berkeley wrote, had the course of empire moved from England to New England, and now the course of an American empire continued westward from New England to the Lake States. The southern lands of Wisconsin, Michigan, and Minnesota were prairie or oak savanna. Their northern lands were boreal forest, the western expression of the Northern Forest that clothed much of New England.[4]

More closely, Michigan, Wisconsin, and Minnesota were where New York, the Empire State, and Maine, the premier logging state, extended their patterns of settlement. They were places and times of extravagant, feckless, promiscuous, and abusive logging, which made them places and times of extravagant, feckless, promiscuous, abusive, and lethal burning. What the axe scalped, the torch incinerated. What the ice had wrought on bedrock and watershed, fire and axe did to forest and wetland.

They violently remade the landscape in ways that would echo through generations.

Once the Great Lakes opened to transportation, the process began, relying on waterways and spring freshets to carry logs cut over the winter. It was the pattern that had prevailed in Maine and much of eastern Canada; it was the same pattern that, at the same time, was felling forests around the Baltic Sea. The world woodland was being cut much as whales were hunted and bison slaughtered to the edge of extinction. This was not folk clearing, but industrial felling. There was little to stay the woodsman's hand. The perception—the belief—that the woods rolled endlessly onward made qualms about exhausting the stock and notions of conservation quaint. There was always more beyond the next height of land. There was little to stop them.

After the Civil War, the pace quickened. Steam opened up the interior of the Northern Woods. The volume of logs picked up; the slash left in their wake was staggering. As soon as one plot was cut out, companies moved on to another. Cut out and get out. They passed over the conifer forests like locusts over fields of wheat. Bad as the era of axe and fire was in New England, it was worse around the Great Lakes. The forest was larger, the climate more continental, lightning more potent, the onslaught of axe and flame more vicious and sudden, the counterbalances fewer and less sticky. The Great Lakes forest had a natural fire regime more vigorous than New England's, and featured a broad transition biome from savannas. Minnesota was almost as large as New York and Maine combined. It had a river and lakes network for moving logs that made Maine's seem quaint. There were no checks or balances, either social or governmental. There was no federal presence willing to intervene in the name of conservation. The roll call of American conflagrations bulges obscenely during the era of Lake State burns, with spikes at 1871, 1881, 1894, 1908, 1910, 1911, and 1918. It was a Gilded Age, an era of robber barons and untrammeled capital, a time that V. L. Parrington later characterized with a macabrely apt phrase as "the Great Barbecue."

The companies insisted cynically they were performing a public service by opening the dense woods to farmers; that they were bringing the capitalism of civilization to the benighted frontier; that they had to cut out and get out to stay ahead of the fires that gorged on moraines of slash. "If it [the North Woods] is to be saved," the *Lumberman's Gazette* argued, "it must be cut as fast as possible." Once begun, there was no pause possible; fires would take what the axe didn't. It was after big burns, the Detroit *Post* insisted, that settlers faced ideal conditions. "These lands offer the best inducement for new settlers. These lands are now in such a condition that they are all ready for seeding to wheat, merely requiring the harrow to be used upon them, in case there is not time to plow." The fire had completed the onerous work of clearing; it had fumigated and fertilized; it had driven off pests and vermin; it had banned the near prospect of future fires since the land was incinerated. Instead the land was ready for occupancy, and high wages were assured since everything would need to be rebuilt. There was even a small trade in memorabilia and trinkets. There was no reason to delay.[5]

Ironically, the primary check was the blowback caused by fires.

Here lies a paradox created when fire ecology met fire history. Free-burning fire was elemental in the dynamic of the boreal forest. Most white pine stands grew on old burns, nearly all birch and aspen followed flames. Fire had coevolved with the Great Lakes forest. What changed was not a sudden immersion of flame into a landscape that had rarely known it, but an orders-of-magnitude increase in its breadth and intensity. The coevolving, check and balance of forest and fire broke down. There was more to burn and less to halt it because the logging companies did not pick up after themselves. The massive debris fields they left behind fed massive fires. The track of westering Americans was a trail of burning, not just figuratively but literally, a long swell that had its tempests and tsunamis, and these overwhelmingly crowded around the Great Lakes. From the 1860s to the 1930s, overlapping the wildland fires that became notorious in the Northern Rockies, these were the Great Fires that gave state-sponsored conservation its lurid pulpit.

Institutions to cope with them were feeble, where they existed at all. The Lake States came halfway in the national narrative, as they stood halfway across the country. They came between private exploitation and public protection. They could look east to New England for one kind of state response, and west to another, more federal model. The political economy of the time favored development, which meant privatization, unchecked capital, and the rapid conversion of wildland to farms. But the newcomers had no knowledge of local landscapes or long-acquired folklore to know where to cut and burn and where to leave well enough alone. They left that knowledge in Europe or the backwoods of New England. Only luck decided whether fires burned hugely or died out in plowed fields and cold swamps.

The wreckage of woods and towns and lives did not go unrecorded. In 1871, the Black Year, as the *Chicago Tribune* called it, the immense fires in Wisconsin were paired with the great Chicago fire that burned simultaneously; *Harper's Weekly* printed engravings still popular today. The 1881 Michigan fires pushed the Red Cross into civilian disaster relief for the first time, and the Weather Bureau sent Sgt. William Bailey to inquire into meteorological conditions, leading to the first published use of the term *firestorm*. Newspapers reported the horrible, thrilling, lurid accounts of folks fleeing, of settlers rescued moments ahead of the flames by trains, of desperate flights into rivers, lakes, and marshes. Most farms failed, and left trashed landscapes, but the sagas end with the trees felled and the soil plowed. The ecological and economic wastage was slower, less visible, less amenable to soaring narratives of heroes and villains. But fires were, and so fires gave light and power to the story.[6]

That outburst of flame inspired one of the two literary traditions of settlement fire (the other being the prairies). That literature spoke not only to fires that were more savage than those to the south, but to people, educated, literate, outraged, and willing to document where the saga of American pioneering had turned to the dark side. We don't know all the big fires of the era, but we know those that became great because the literate class—priests, schoolteachers, ministers, journalists—wrote down what they had witnessed and often photographed them, like ghastly battlefields in a misguided war on nature. No fire has inspired so many books as the 1871 Peshtigo burn. New books continue to be written for Hinckley, for Baudette, for Cloquet. No other region, save perhaps Southern California in the post–World War II era, has so rich a documentary record. The first enduring images of forest fires in America are those that illustrated the grotesque bestiary of flame loosed and gorged on the overdosed combustibles around the Great Lakes. The memory still lingers in the 21st century in Jim Harrison's *True North* (2005) where the young scion of an old timber baron who had "laid waste" to the Upper Peninsula marvels that "the grandeur of the destruction had been mythologized in story and song" before he seeks to purge that past from his future.[7] That institutions did emerge is a tribute to those who vividly chronicled the ruin and terror.

These were frontier fires, a part of the inevitable violence (in this case environmental) by which the land and its occupants changed. All parties recognized this fact, and they all appreciated that the fires would fade away as the frontier passed. Once converted from wildland to farm and town, the big burns would vanish from the land. The wild would be domesticated. Slash would become fallow, and wildfires, field fires. The issue was whether the ride would be a waterfall or a rapid, whether society, specifically, the state, should intervene to quiet the axe and quench the torch and extend the process in a less ferocious form or whether it should let frontier pass as rapidly as possible. A few conflagrations, like the occasional stock market crash, were the price of progress. The Great Barbecue had flames before it had coals.[8]

This was the Northeast saga on steroids. What happened over two centuries in New England occurred in a few decades in the Lake States, occurred in an environment more prone to explosive burning, occurred without the folk mores and local traditions that, here and there, had braked the process even in Maine or the state interests that had in New York led to the creation of forest preserves. So, too, the Lake States looked to the Northeast for inspiration in quelling the havoc, again transitional in its juggling of state and federal actions. And so, also, its lands, like those of New England, underwent a cycle of abandonment and recovery.

With the big burns as backdrop, forest conservation became an issue as the Lake States began to organize state forest bureaus and acquire forest lands for the state to protect or reforest. Forest commissions were impaneled, forestry schools founded, and forest bureaus authorized. The project became a major political movement during the administration of Teddy Roosevelt. In 1905 he transferred the existing forest reserves from the Interior Department to the Forest Service in the Department of Agriculture. In 1907 he doubled the size of the national forests. In

1908 he hosted the famous Conference of Governors on conservation at the White House; the next year, those ambitions swelled into the North American Conservation Conference. Continued by his successor, President Taft, the first national forests for Minnesota and Michigan were created in 1908 and 1909. The next year, the fabled Year of the Fires, the Lake States Forest Fire Conference was held in St. Paul—the first such gathering in American history. By now serious timber companies recognized that they could not continue amid the threat of wildfires; in Wisconsin and Michigan they agreed to pool efforts into a Northern Forest Protective Organization. Industry, the states, and the federal government would have to collaborate or flame would take the last stands and prevent any hope of regeneration. In 1911 the Weeks Act established a mechanism for federal-state cooperation, prompting a surge of state forestry bureaus. The more expansive Clarke-McNary Act in 1924 broadened the range of support. The final surge of federal interest came during the New Deal, when tax-delinquent lands became targets for acquisition and the CCC provided a workforce to fight fire and replant woods.[9]

Meanwhile, the amount of cutover, burned-over, and abandoned lands spread like cancer. Such lands were not merely unproductive: they were a menace, a point of pyric infection. They would not mature into rich farmlands—that would happen on the former savannas and prairie to the south; and without formal protection, particularly from fire, they would never regenerate into serviceable woods. Instead the states inherited vast swathes of cutover that owners walked away from rather than pay taxes on stumps and ash pits. These became the nucleus for state forests, as much on the model of the Adirondacks and Catskills as on the newly gazetted Superior, Huron, and Marquette National Forests. Wisconsin lagged. In 1910 voters had approved purchase (at fire-sale prices, as it were) for state forests, but then developers pushed back, and in 1915 the Wisconsin Supreme Court voided the act. Not until 1924 did citizens effectively reinstate it.

Still, there were new lands yet to cut, and the axe kept up its work. When the prime white pine was gone, loggers started in on jack pine and aspen, many of which had sprouted in the ruins, and switched from saw timber to pulp, like miners sluicing through tailings, or hunters gathering the bones of bison for fertilizer. But finally too many lands were stripped, and even the organic soil burned away. The communities, too, burned off, or rotted away. The only strategy left was abandonment. The axe went to the still-extant big trees of the Northwest and the Southeast. The two years after Clarke-McNary were bad for fires but worse, economically, for farming. When drought and Depression struck in the 1930s, opposition collapsed. In 1931 nearly a million acres burned in Minnesota alone. The old mirage of converting forest to farm evaporated. Farms would stay south, forests would return north, and this time the state would keep fires out of both. However the future might evolve, it would include fire protection.

After the Laurentide ice had left, as its vast burden melted away, the land had slowly rebounded upward. So when the heavy hand of settlement lifted, the biotic landscape began to rise from the ruins. The land went into rehab. As in New England the forest, much of it a different forest, returned.

The oracle of the New England countryside, Henry Thoreau, found a Lake States echo in Aldo Leopold, and *Walden* with *A Sand County Almanac*. Leopold's shack, like Thoreau's cabin, has become a tourist destination, if not a pilgrimage. The land ethic has become as vital for modern environmentalism as civil disobedience for the civil rights movement. When Leopold acquired the worn-out land that became the farm, with the chicken coop that morphed into the shack, the Lake States landscape was probably at low ebb. Not content to argue only for preservation in the guise of wilderness, Leopold also worked at restoration. The land ethic applied to both. The tools that had unraveled landscapes could, with intelligent care, begin to wind them back together.

His personal project coincided with wholesale government efforts to the same end. State and federal agencies acquired more abandoned land. The Soil Conservation Service stabilized soils. The Resettlement Administration helped the relocation of farms from the woods to the prairies. The CCC planted trees, reflooded some wetlands, erected lookout towers, and fought fires. The infrastructure for settlement segued into one for reconstruction. As land passed from laissez felling and firing to organized agencies, fire seeped away. There were fewer people committed to starting fires and more empowered to putting them out. After the war years, as the Forest Service became a conduit for surplus military equipment, the region established an equipment development center at Roscommon to assist the conversion from front lines to firelines. The economy shifted from raw commodities to services, particularly recreation. Big burns ceased to be routine. The Great Fires remained seared into collective memory, and celebrated in museums, but nothing like the community-consuming megafires returned. When the Forest Service sought to promote a laboratory for forest fire research, as it was doing in the Southeast, Northern Rockies, and Southern California, there was little enthusiasm, and the idea withered away. The problem fires were a bad nightmare from the past. The national narrative for fire, like the old loggers, had moved on.

Revealingly, Aldo Leopold, then at the shack, died on April 21, 1948, when he spotted a fire on a neighbor's land. It had started in trash, the domesticated slash of a tamer time. A flank of the fire moved toward marshlands and flashed toward a patch of pines he had planted. Neighbors gathered to save the neighbor's house, while Leopold took a backpack pump and worked the errant front. He collapsed with a heart attack. The flames burned lightly over him and inscribed the final entry into his notebooks with scorch marks. The next year *A Sand County Almanac* was published posthumously.[10]

His death by fire came midway between the holocausts of the past and the problematic burns of the future. Thirty years had passed since the last monster fire had threatened Cloquet. Twenty-eight more years would pass before bad fires again returned to Seney.

The region's fires did not suddenly cease. They remained common, but as with the Northeast, they had few opportunities to blow up and run freely through trashed landscapes. Wisconsin, Michigan, and Minnesota created prominent bureaus to manage the states' public forests and formidable fire control apparatus to throttle fire out of the rural landscape. The biotic partition of logging from farming quelled the most provocative causes. But the old boosterism that had invited new settlers to seize the burned lands because the fire threat was over had, in macabre ways, proved true. The new regimes kept bad fires small. The region banished its fires into memorial plaques and county museums, while the country boxed up that awful heritage and put it into archives.

The land healed. The forests came back. Large swathes were now public land, either state or federal, and these felt the impress of an environmental movement that bubbled out of the 1960s and was codified in the 1970s. Public lands began to specialize, spinning off wildlife refuges, national parks, and wilderness—the most celebrated, at Boundary Waters, not far from the disaster fires of 1908 and 1910. Fires returned. They were modest by historic standards, like fires smoldering in peat that pop up when the winter snow melts off. But institutions, however reformed, were in place this time, not the wistful hope of reformers. Scientists probed the past for fire scars and charcoal in lake varves, extending the region's fire history beyond the eruptive era. Fire officers were less obsessed with suppressing big burns, which seemed unlikely, than with restoring good fires.

The best known wildfires were, in fact, escapes from prescribed burns— the core emblem of the national fire revolution—gone bad on public lands. Two in particular frame the 1970s. In mid-May 1971 a holdover from a prescribed burn blew up on the Superior National Forest. The Little Sioux fire blasted across 14,628 acres and stunned fire managers not accustomed to fires of this size, even if it was two orders of magnitude smaller than the historic burns. The decade closed with an escaped prescribed fire on the Huron National Forest to promote habitat for the endangered Kirtland's warbler. Some 30 hours and 24,000 acres later, the fire had burned 44 homes and structures in the hamlet of Mack Lake and killed one firefighter. Both fires led to model case studies by Forest Service fire researchers. Both caught national attention. But neither was likely to stake claims that the region was again in the vanguard of national fire reforms.[11]

Rather, the episodes appeared as cautionary tales, of what can happen when a region that had (to national interests) lost its fire capacity, had managed to dampen rural burning and to quickly kill wildfires, tried to adopt cutting-edge practices for a new era. Rightly or wrongly, the Lake States remained on the periphery. Their most active fire management lay in restored prairies, not a recovered North Woods.

There seem no unique fire problems that the rest of the country looked to the Lake States to resolve. The fires that preoccupy the region are local expressions of national themes—wilderness, WUI, prescribed burning, fatalities. The Lake States is the critical habitat for none of them. The core region for prescribed fire is Florida. The core for backcountry burning is the Northern Rockies. The core for wildland-urban fringe fires is California. What the Lake States bring to the fireline is history.

———

After the wave of slashing and burning that swept over Seney, developers moved in to encourage farming. A land company cut ditches to drain wetlands and swamps. Farmers plowed and burned, and then, as their forebearers had in New England, left. Better to abandon than pay taxes on trashed land. The lands fell to the state, which in the case of Seney wanted to off-load the burden onto the federal government by proposing the feds convert it into a wildlife refuge.[12]

Amid the New Deal, conservation programs flourished. The upshot was Seney National Wildlife Refuge. The Works Progress Administration and Civilian Conservation Corps moved in and began systematically rehabilitating what earlier visitors had systematically wrecked. The wetlands reflooded. The woods grew back. Wildlife, particularly migratory birds, returned. And so, too, slowly, probingly, feeling its way back like black bears and badgers, did fire.

Fire had been away a long time. The lavish slash that had fueled its riotous explosions was gone. Loggers, farmers, promoters, transients—all those who had set those old fires had left. Fire protection had spread over the peninsula, and in the postwar years, mechanized. Some 25,150 refuge acres went into legal wilderness. It would take time for spark, woods, and weather to recombine. No one lamented the lost conflagrations, but wildlife specialists did worry about the complete remission of burning. Some fires would happen, and some fires were useful, some were the emergent properties of new opportunities on old landscapes. Eventually, fire would return in avatars, wild and prescribed, that suited its new circumstances.

In 1976 they all came back, all at the same time, all in the same place. Seney had not suffered a major drought in 40 years or a significant fire for 70. Outside the refuge, a wildfire started on the Lake Superior State Forest. Within, on July 7 the refuge staff kindled the Pine Creek prescribed fire, an experiment in habitat maintenance, originally planned for 40 acres but, after Department of Natural Resources protests, shrank to one acre, which smoldered stubbornly in dry muck. (Intermittently since the 1940s, with uncertain results, the refuge had tried some controlled burns.) On July 30 lightning started a fire, later named Walsh Ditch and treated as a prescribed natural fire, in a recently designated wilderness area. Three categories of land, three management goals, three fires.[13]

The Michigan Department of Natural Resources suppressed, with mechanized equipment, the wildfire. The Pine Creek prescribed fire slowly burned on, gnawing through dense regeneration and peat until, with the assistance of the DNR, the refuge stalled the fire at 200 acres. The Walsh Ditch fire, monitored intermittently by air, swelled to 1,200 acres and seemed to gather strength for a rush outside its sanctuary. The refuge sought help from the DNR to contain it while also requesting national assistance in what escalated into the largest mobilization of the year and became the most expensive fire in Department of the Interior history to that time. The imported fire team elected to burn out the entire wilderness area. Eventually, the Pine Creek fire was drowned after diverting a stream with bulldozers. On September 7 the Walsh Ditch fire was declared controlled. But the drought deepened, the smoldering burn found fresh fuels, the fire blew up and out of the refuge. Arson fires appeared. The demobbed crews were sent back. When snow finally quenched the flames, the Walsh Ditch fire had blackened 72,500 acres. Fish and Wildlife Service reviews suggested that the ecological effects of the fires have, overall, been positive.

The experience of 1976 was deeply inconclusive. Fire managers knew they could not simply suppress every fire, and they knew, too, that many landscapes needed the right kind of fire. Yet history was an unreliable guide—what kind of restored fire suited this kind of recovered land? Ecology understood the basic principles, but fire is a particular event at a particular place, and it was unclear what the outcomes would be. (They were mostly surprising.) The cost of the outbreak was obscene, not only in money but in political attention. With several other stumblings to follow, the fires put the Fish and Wildlife Service on the path to a national reformation of its program. Revealingly, the catalytic event occurred in Florida, not the Lake States.[14]

Still, the summer's fires were a cameo for the region. With the recovered biota had come a recovery of fire. In New England the public lands that serve as a prime habitat for free-burning fire were small and scattered, and outside sandy soils and swathes of true boreal woods, not inherently fire prone. In the Lake States they were large enough to support big burns, and there were opportunities for fires in the spring and fall of most years, and increasingly summers. The historic burns had gorged on the carcass of scalped lands. The new fires had to incorporate ecological goods and services: the birder's binoculars, the ecologist's transect, the rod and gun of hunter and fisherman, the open space of the recreationist.

No one today looks to the Lake States for insights into the critical fire issues of the day. But the region holds important lessons about how to manage fire in rehabilitated lands, about fire management collectives where the states have equal or greater powers than federal agencies, about the character of the workforce that contemporary fire management requires, about what kinds of fire are appropriate. To date, efforts to import intact the systems of the Southeast, Northern Rockies, or California have stumbled. The Lake States require their own solutions.

Yet in an era that promises a future of fusion fires, of hybrids between simple suppression and burnouts that resemble prescribed fires, the region might enjoy another renaissance because it is, historically and geographically, a transition zone. The Lake States stand midpoint in America's pyrogeography—geographically, between Northeast's private land tenure and the Northwest's heavily public lands; historically, between the era of laissez-faire cut-out-and-get-out and state-sponsored preservation. The Lake States have pieces that exist (and have purer expression) elsewhere. What the region features uniquely is a chronicle of how those pieces have come together.

In times of uncertainty and rapid change, the generalists survive. The variety that prevents the Lake States from imposing a model on the rest of the country may prove the region's greatest asset in a future that promises changes as radical in outcomes, if more subtle in means, than occurred in the settlement past. The nation's fire strategists could do worse than take a few days and follow Nick Adams into the fire equivalent of the Big Two-Hearted River.

EPILOGUE

The Northeast Between Two Fires

MUCH AS IT CONCENTRATES storm tracks, funneling them between the Great Lakes and the Gulf Stream, so the Northeast has distilled histories, both natural and human, passing them between fires. Surface fires and subterranean fires. Big fires in the past, small fires today. Fires that burn living landscapes, fires that burn lithic ones. Fires from nature, fires from people. Fires that once informed the national narrative, fires that lie outside the main current of contemporary fire, that spin like so many eddies by a sand bar or sit like oxbow lakes left from the past. Fires scattered like tiny stars in search of a constellation to organize them.

More than other fire regions the Northeast has a mix in which no single principle seems to dominate. The mix itself is the story. Its fires reside within that churn of weather, land use, disturbances, and historical eras that pass like the seasons. No other region has such a medley of fires and fire eras. Here fire is complex because its setting is complex, and its setting is complex because, for so long, it has been shaped by the peculiar interactions of particular peoples and localized natures. It has lightning fires, but not enough to inform a fire-dependent biota. It has anthropogenic fire, but those fires cannot shape the scene without assistance from axe, plow, or livestock. Its pyrogeography, like its land tenure, is a palimpsest. It's impossible to look at interior Alaska or the longleaf coastal plains or California shrublands and not see fire. It's easy to overlook fire in the Northeast because there is so much else going on. Until, at certain times, the fires become undeniable and channel their flames through institutions and spread far and wide.

The Northeast has had its cycles of Great Fires. They influenced the national story as profoundly as the Great Fires of the Rockies, the prescribed fires of Florida, or the firefights along California's South Coast. Out of the Northeast came the grand prophets of conservation, from George Perkins Marsh to Gifford Pinchot. The fires they denounced in their jeremiads were the fires they saw in the wracked lands being slashed and burned around them. That other Great Fire, first kindled and fueled in the Northeast, industrial combustion, continues to act like a software operating system on all the other fire regimes in the country. That it is so often invisible can stand as a symbol both of what the Northeast has contributed and of why it is so easily overlooked.

Amid today's enthusiasms for restoring fire at a landscape scale it is easy to dismiss the Northeast as malign in its history and trivial in its attempts to atone. The East misread the fire scenes of the West and the Southeast, insisting that fire had no natural or legitimate purpose. So, today, those regions may reciprocate by insisting that fire must be present in abundance for land management to be authentic and integral. But eras pass; blowback isn't policy. If we are to understand fire's ecology, we must understand it everywhere. Getting fire to scale can mean getting it right small as well as large.

So, too, it is fashionable today to condemn the technological power unleashed by industrial combustion as underwriting the manifold insults of the Anthropocene. Yet history's wheel may turn again, and we may come to see today's era as a critical transition, an experiment in humanity's fire management that paradoxically stalled the return of the ice and that segued to an age of combustion-free energy. The ice is overdue. It may be that once again our firepower will save us as much as it threatens us. More than other regions, the Northeast, and its Lake States extension, all vulnerable to continental-scale ice sheets, may know the positive consequences as well as the ecological compromises, and it may pioneer, once again, the social arrangements needed to temper disruption into resilience.[1]

The national story has mostly tracked fire on our common lands, the public domain, overwhelmingly in the West, lands that can pretend to hold the contact landscapes we call wilderness. We are less drawn to stories in which cultural landscapes pile on one another, in which fire scars overlay one another like jackstrawed lodgepole, in which the nominal clarity of a choice to put fire in or take it out is muddled by places that represent a deep interaction between nature and culture, and between fire and a score of other disturbances, and in which we can't pretend otherwise. The region's pyrogeography consists of ecological overlays, articulated in four dimensions. Its narrative nearly trips over subplots and plot twists.

Fire history resembles a churn of fire that follows a churn of biota. Old species have gone, new species have arrived. Old regimes have faded, new regimes have come into focus. Yet only in places is fire obligatory. If you keep fire out of most western or southeastern settings, the conditions change in ways that make fire, when it returns, more severe. If you keep fire out of most northeastern settings, the landscape will morph into more fire-immune forms. What does fire mean? At what cost should fire be restored?

Across the country the pressures against reintroducing landscape fire are almost everywhere daunting. In some places like Southern California they can assume grandiose, even megalomaniacal proportions. In the Northeast they more resemble a swarm of black flies, biting away at flame with dense road networks, air quality concerns, a too-brief burning season, the relative lack of leverage from endangered species, the absence of a deep-rooted fire culture, the tenacity of local rule with its high demands on social capital, and the scattershot distribution of fire-dependent habitat. Add to that swarm the fact that big fires have not traumatized or frightened the region since 1963 or 1947. The impression is that fire management is not mandatory and that, in many places, neither is fire suppression.

Those who train here go West to satisfy their hours and qualifications or to conduct their research. The Nature Conservancy defunded its global fire initiative. The National Park Service has downsized its operations. Wildland Restoration International, a company developed to service prescribed fire needs for the region, mostly looks outside it to Belize, Tanzania, and the Bahamas to pay its bills. When Bill Patterson retires completely, there is no expectation that UMass will replace him with a fire scientist.

Yet the Northeast has circumstances that make it relevant. Fire ecology must explain its fires as well as those on Colorado's Front Range or the Mogollon Rim. More, its fire regimes are unblinkingly bound to people, and there is no pretending otherwise. The assumption of many planners, especially those trained in American economics, is that a "public option" is a "nonstarter." But the Adirondacks, at six million acres the largest park in the continental United States, is a mix of public and private lands under a common agency. The New Jersey Pinelands regulates development over 1.1 million acres of mixed ownership. A public corporation oversees the Albany Pine Bush Preserve. The Manichaean division of land into either all public or all private makes little sense in the Northeast, which has crafted working alternatives. Besides, it's worth recalling that the great public reserves in the West were the product of New England minds working the political system of the Gilded Age. There are lessons to be learned for the West.

<hr />

Its churn of landscape change is what makes the Northeast compelling. It shows how profoundly fire history can feature cycles compounding cycles, with fire expressing itself as circumstances permit. The Northeast challenges our notion that wilderness is the first and best example of nature protection and that institutions only respond to fire, not shape it. It favors a negotiated fire regime, not a restored natural one.

Three times the national fire narrative has begun here. The Northeast was the major portal for European notions of fire and fire practices, for ideas about fire and conservation that found full play in the West, and for the entry of industrial combustion. Now the Northeast is scouting a trail for different kinds of land tenure, not solely either public or private, but with collaborations and easements that mingle the two, and in which reclamation is partial. These arrangements will determine what kinds of fires are possible and what forms fire management will take.

The Northeast's fires will assume the properties of their settings. Its fires will not be singular engines of ecology, but will be embedded within a swirl of disturbances and processes, mostly from the hands of humans. It's a more prosaic, less epic view of fire. But it's a perspective that will likely become more common and vital as the United States continues to settle and unsettle its landed estate in ways that are difficult to disentangle or parse into stand-alone pieces. A fire region often dismissed as burdened by or lost in its past may show a different kind of future.

The Northeast has repeatedly influenced the national narrative. There is no reason to think it won't do so again.

NOTE ON SOURCES

MOSTLY I HAVE RELIED, as with my other reconnaissances, on site visits, talks with hosts, and documents from agencies and journals. The individual essays cite those works I found particularly useful.

But the Northeast as a whole has a spectacular cache of recent fire studies and an unusually rich historiography to go with its long chronicle of European contact. The cache is the product of a stress-test survey commissioned by the Northeastern States' Forest Fire Protection Compact and written by the Irland Group. Its output includes statistical summaries of fire for each of the compact's members, the Sixties drought and fire, trends, and the capacity of the present NFFPC to meet possible fire challenges. It's a remarkable collection and the obvious portal for anyone interested in the contemporary scene. The documents can be accessed through the North Atlantic Fire Science Exchange website, which also hosts a wonderful collection of recent research.

For background into regional history, I found particularly helpful Richard W. Judd's *Second Nature: An Environmental History of New England*, Lloyd C. Irland's *The Northeast's Changing Forest*, and for its value as a historical baseline Charles Sargent's *Report on the Forests of North America (Exclusive of Mexico)*, his contribution to the 1880 census. For the deeper history, I referred to George P. Nicholas, ed., *Holocene Human Ecology in Northeastern North America*. To help sort out the background stories of the many states, I appealed to Ralph R. Widner, ed., *Forests and Forestry in the*

American States: A Reference Anthology, a breezy, popular survey without further sources but helpful handholds on the cliff for someone new to the region. For the Lake States I continue to find Susan L. Flader, ed., *The Great Lakes Forest: An Environmental and Social History*, a pertinent point of departure. For a survey of the critical oak forests, I relied on Matthew B. Dickinson, ed., *Fire in Eastern Oak Forests: Delivering Science to Land Managers*, General Technical Report NRS-P-1, U.S. Forest Service, 2006, to introduce me to the major themes and researchers.

NOTES

PROLOGUE

1. Quoted in Timothy Dwight, *Travels in New England and New York*, ed. Barbara Miller Solomon, vol. 3 (Cambridge: Harvard University Press, 1969), 350–51.

2. See Stephen Pyne, *Fire in America: A Cultural History of Wildland and Rural Fire* (Seattle: University of Washington Press, 1995), 56, and Stephen Pyne, *Awful Splendour: A Fire History of Canada* (Vancouver: University of British Columbia Press, 2007), 124–27. The standard popular reference has been David M. Ludlum, "New England's Dark Day: 19 May 1780," *Weatherwise*, June 1972, 112–19, quote on page 113.

    The 1780 Dark Day has been reexamined in the light of fire scar data by Erin R. McMurry, Michael C. Stambaugh, Richard P. Guyette, and Daniel C. Dey in "Fire Scars Reveal Source of New England's 1780 Dark Day," *International Journal of Wildland Fire* 16 (2007): 266–70, which includes a thorough update of sources.

A SONG OF ICE AND FIRE — AND ICE

1. ICE as an acronym comes from Alfred Crosby, *Children of the Sun: A History of Humanity's Unappeasable Appetite for Energy* (New York: Norton, 2006).

2. For a tidy summary, see William A. Patterson III and Kenneth E. Sassaman, "Indian Fires in the Prehistory of New England," in *Holocene Human Ecology in Northeastern North America*, ed. George P. Nicholas (New York: Plenum Press, 1988), 107–36. Also useful are Carl Sauer's historical surveys,

though New England is a minor constituent: *Sixteenth Century North America: The Land and People as Seen by Europeans* (Berkeley: University of California Press, 1971) and *Seventeenth Century North America: French and Spanish Accounts* (Berkeley, Calif.: Turtle Island Foundation, 1977).

3. William Wood, *New England's Prospects* (1634; repr., Boston: Prince Society, 1865), 17.

4. Thomas Morton, *New English Canaan* (1637; repr., New York: Arno Press, 1972), 52–53.

5. Morton, 53–54; Van der Donck quoted in Gordon M. Day, "The Indian as an Ecological Factor in the Northeastern Forest," *Ecology* 34, no. 2 (1953): 337.

6. Peter Kalm, *Travels into North America*, trans. John Reinhold Forster (1770–71; repr., Barre, Mass.: Imprint Society, 1972), 210, 361.

7. Timothy Dwight, *Travels in New England and New York*, ed. Barbara Miller Solomon (Cambridge, Mass.: Harvard University Press, 1969), 4:37–39.

8. My Thoreau observations derive from David Foster's very congenial and helpful *Thoreau's Country: Journey Through a Transformed Landscape* (Cambridge, Mass.: Harvard University Press, 1999); block quote from page 10.

9. Quotes from Foster, 116–17.

10. Quotes from Foster, 120–21.

11. Figures from Mark B. Lapping, "Toward a Working Rural Landscape," in *New England Prospects: Critical Choices in a Time of Change*, ed. Carl H. Reidel (Lebanon, N.H.: University Press of New England, 1982), 61.

12. Charles S. Sargent, "The Protection of Forests," *North American Review* 135 (1882): 397, 392; Charles S. Sargent, "Protection of the Forests of the Commonwealth," Massachusetts Horticultural Society, February 13, 1880, published note from secretary.

13. Statistics from Lloyd C. Irland, "Massachusetts Fire History Working Paper: Revised Draft, May 24, 2012," Irland, "Vermont's Fire History 1905–2011: Initial Observations, with Analysis of Individual Fire Data, 1977–2011," and Irland, "Fire History of New York State, 1920–2010: Working Paper," all available through the North Atlantic Fire Science Exchange website: http://www.firesciencenorthatlantic.org/search-results/?category=Technical+Report. See also Julie A. Richburg and William A. Patterson III, "Fire History of the White and Green Mountain National Forests: A Report Submitted to the White Mountain National Forest, USDA Forest Service, January 31, 2000." The site is a mother lode of regional fire information, with histories for other member states as well.

14. Lloyd C. Irland, "The Northeast's Great Sixties Drought: The Fire Outbreak," Northeastern Forest Fire Compact, draft work paper, at the North Atlantic Fire Science Exchange website. The 1947 fires have been the subject

of numerous articles, three documentary films, and one popular book, Joyce Butler, *Wildfire Loose: The Week Maine Burned* (Kennebunkport, Maine: Durrell Publications, n.d.).

15. David B. Kittredge, "The Fire in the East," *Journal of Forestry*, April/May 2009, 162–63.

WHERE THE PAST IS THE KEY TO THE PRESENT

1. For a wonderful introduction to the place and its themes, see David R. Foster and John D. Aber, eds., *Forests in Time: The Environmental Consequences of 1,000 Years of Change in New England* (New Haven, Conn.: Yale University Press, 2004).

2. Hugh M. Raup, "The View from John Sanderson's Farm: A Perspective for the Use of the Land," *Journal of Forest History* (April 1966): 9. "Anti-planning" quote from Brian Donahue, "Another Look from Sanderson's Farm: A Perspective on New England Environmental History and Conservation," *Environmental History* 12 (January 2007): 19.

3. Donahue, "Another Look," 9.

4. Donahue, 20.

5. David R. Foster, "Three Views from John Sanderson's Woodlot," *Forest History Today*, Spring/Fall 2014, 45.

FIRE'S KEYSTONE STATE

1. I was fortunate in having some wonderful guides in my road trip through Pennsylvania: Ben Jones of the Pennsylvania Game Commission; Mike Kern, chief of the Division of Forest Fire Protection, Department of Conservation and Natural Resources; and Pat McElhenny of the Nature Conservancy's Hauser Nature Center. They were generous with their time as well as with documents. My thanks to them all.

2. For a useful sketch of Pennsylvania as a fire environment, though now somewhat dated, see Donald A. Haines, William A. Main, and Eugene F. McNamara, *Forest Fires in Pennsylvania*, Research Paper NC-158, U.S. Forest Service, 1978. Though ostensibly on oaks, the following conference includes papers that range more widely: Matthew B. Dickinson, ed., *Fire in Eastern Oak Forests: Delivering Science to Land Managers: Proceedings of a Conference*, General Technical Report NRS-P-1, U.S. Forest Service, 2006. See particularly William A. Patterson III, "The Paleoecology of Fire and Oaks in Eastern Forests"; Richard P. Guyette et al., "Fire Scars Reveal Variability and Dynamics of Eastern Fire Regimes"; Charles M.

Ruffner, "Understanding the Evidence for Historical Fire Across Eastern Forests."

The most thorough inquiries have come from Marc Abrams and his colleagues. See, for example, "Fire and the Development of Oak Forests," *BioScience* 42, no. 5 (May 1992): 346–53, and Marc D. Abrams, David A. Orwig, and Thomas E. Demeo, "Dendroecological Analysis of Successional Dynamics for a Presettlement-Origin White-Pine-Mixed-Oak Forest in the Southern Appalachians, USA," *Journal of Ecology* 83 (1995): 123–33.

3. On Pennsylvania bison see E. Douglas Branch, *The Hunting of the Buffalo* (Lincoln: University of Nebraska Press, 1962), 59–62.

4. Joseph M. Marschall et al., "Fire Regimes of Remnant Pitch Pine Communities in the Ridge and Valley Region of Central Pennsylvania, USA," *Forests* 7, no. 224 (2016): doi:10.3390/f7100224.

5. Patrick H. Brose et al., "The Influence of Drought and Humans on the Fire Regimes of Northern Pennsylvania, USA," *Canadian Journal of Forest Research* 43 (2013): 757–67. For a simplified version, richer in the human history, see Patrick H. Brose et al., "Fire History Reflects Human History in the Pine Creek Gorge of North-Central Pennsylvania," *Natural Areas Journal* 35, no. 2 (2015): 214–23. An interesting review of the impact of indigenous peoples is available in Bryan A. Black, Charles M. Ruffner, and Marc D. Abrams, "Native American Influences on the Forest Composition of the Allegheny Plateau, Northwest Pennsylvania," *Canadian Journal of Forest Research* 36 (2006): 1266–75.

6. The best reference is John E. Thompson, "Interrelationships Among Vegetation Dynamics, Fire, Surficial Geology, and Topography of the Southern Pocono Plateau, Monroe County, Pennsylvania" (master's thesis, University of Pennsylvania, 1995). An excellent distillation of current knowledge that also contains a map of potential barrens is available through Pennsylvania Game Commission, "Barrens Habitat," available online at http://www.pgc.pa.gov/Wildlife/HabitatManagement/Documents/Barrens_Chapter.pdf.

7. Ralph C. Hawley, "Forest Fires in the Poconos," *Forest Leaves* 23 (1933): 157.

8. See Terry Jordan and Matti Kaups, *The American Backwoods Frontier: An Ethnic and Ecological Interpretation* (Baltimore, Md.: Johns Hopkins University Press, 1992).

9. Frame quote from Clarence Glacken, *Traces on the Rhodian Shore* (Berkeley: University of California Press, 1967), 697. Harris quote from Thaddeus Harris, "The Journal of a Tour into the Territory Northwest of the Alleghany Mountains, Made in the Spring of the Year 1803," in *Early Western Travels, 1748–1846*, ed. Reuben Gold Thwaites, vol. 3 (Cleveland, Ohio: A. H. Clark,

1904), 327. My observations in this paragraph paraphrase those made in Stephen Pyne, *Fire in America* (Princeton, N.J.: Princeton University Press, 1982), 52–53.

10. Charles S. Sargent, "The Protection of Forests," *North American Review*, 1882, 390.

11. Charles Sprague Sargent, *Report on the Forests of North America (Exclusive of Mexico)* (Washington, D.C.: Government Printing Office, 1884), 510.

12. Brose et al., "Influence of Drought and Humans," 765–766.

13. Sargent, *Report on the Forests of North America*, 491–92, 506.

14. Lester A. DeCoster, *The Legacy of Penn's Woods, 1895–1995: A History of the Pennsylvania Bureau of Forestry* (Harrisburg: Pennsylvania Department of Conservation and Natural Resources, Bureau of Forestry, 1995), 11.

15. J. T. Rothrock, *Preliminary Report on Forest Fires* (Harrisburg, Pa.: Busch, 1896), 4. On Pennsylvania's fire laws, see DeCoster, *Legacy*, 16. Gifford Pinchot, "Study of Forest Fires and Wood Protection in Southern New Jersey," *Annual Report of the State Geologist for the Year 1898* (Trenton, N.J.: MacCrellish and Quigley, State Printers, 1898): 11; Pinchot, *Forest Protection: Fire Prevention and Extinction*, Bulletin 27 (Harrisburg, Pa.: Department of Forestry, 1922), 1. George Wirt, *Lessons in Forest Protection*, Bulletin 35 (Harrisburg, Pa.: Department of Forests and Waters, 1927), 17; Gifford Pinchot, *A Primer of Forestry: Part 1* (Washington, D.C.: Government Printing Office, 1899).

16. Quote from Samuel P. Hays, *Wars in the Woods: The Rise of Ecological Forestry in America* (Pittsburgh, Pa.: University of Pittsburgh Press, 2007), 121. Statistics from Bureau of Forestry. Warden numbers from Mike Kern, Bureau of Forestry, interviewed on September 25, 2017.

17. The original records reside with the Pennsylvania Bureau of Forestry, supplemented with materials at the Cameron County Historical Society (note: the Wildland Fire Lessons Learned Center has digitized the collection), hereafter referred to as Pepper Hill Fire Documentation. I relied on the file titled "Pepperhill Fire Order of Events" for reconstructing the sequence of fires and responses.

18. Quotes from William M. King and Frank H. Bender, "Initial Report, File No. 2-D-06–2, October 22, 1938," File 1, Pepper Hill Fire Documentation, 2.

19. King and Bender, 2.

20. Quote from Cameron County Historical Society, "The Pepper Hill Fire and Memorial Springs." Kammrath's name is spelled variously in the reports, also as *Kamarath* and *Kammarath*.

21. There are many versions; I use the coroner's report: Dr. J. D. Johnston, "Inquisition Indented and Taken at Emporium, in the County of Cameron,

State of Pennsylvania, Beginning on the 31st Day of October, A.D., 1938," File 5, Pepper Hill Fire Documentation, 3.

22. The most accessible source for the quotes is Wildland Fire Lessons Learned Center, "Overview of the Pepper Hill Fire of 1938," 4–6.

23. Quotes from "Supplemental Report, File No. 2-D-06–2, October 24, 1938," Pepper Hill Fire Documentation.

24. Size from official Bureau of Forests and Waters fire report, File 9, and cost from File 7 in Pepper Hill Fire Documentation.

25. File 5, Pepper Hill Fire Documentation.

26. Pvt. W. M. King, "Confidential Report: Record of the Investigation Conducted into the Deaths of Seven Members of C.C.C. Camp S-132 in a Forest Fire in the Lick Run District of Cameron County, Which Occurred October 19, 1938, by a Board of U.S. Army Reserve Officers Convened for This Purpose," 4. File 3, Pepper Hill Fire Documentation. The testimonies of those on the fire is frequently riveting.

27. File 8, Pepper Hill Fire Documentation.

28. File G-Letter, Pepper Hill Fire Documentation.

29. There are several competing versions of how the fire started, all of which shift the blame. I have selected the variant told in David DeKok, *Unseen Danger: A Tragedy of People, Government, and the Centralia Mine Fire* (Philadelphia: University of Pennsylvania Press, 1986) as the most credible.

30. See Christopher F. Jones, *Routes of Power: Energy and Modern America* (Cambridge, Mass.: Harvard University Press, 2014), which is especially good at tracing the social and political interplay between new sources of power and their application.

31. A. A. Brown and A. D. Folweiler, *Fire in the Forests of the United States* (St. Louis, Mo.: John S. Swift, 1953), 3. The original volume, by Folweiler, was published in 1937; Brown revised it, and the quote is the opening paragraph of the 1953 edition, and is in his words.

32. Hays, *Wars in the Woods*, 119.

33. Gifford Pinchot, *Breaking New Ground* (Covelo, Calif.: Island Press, 1998), 137.

BOG AND BURN

1. Lee Brothers Inc. have 217 acres of cranberry bogs, and 1,650 acres of forest, a little under an 8:1 ratio rather than the ideal 10:1.

I have long known about New Jersey's pyric Pinelands, but not until Bob Williams, a consulting forester, offered his services as a guide did I have the opportunity to see for myself. Over the course of a day and a half, he provided a crash tutorial. Others who contributed to the conversation include Bill Edwards, Shawn Judy, Steven Lee (father and son), Horace Somes, Jim Dusha, Bill Zipse, Ken Clark, Mike Gallagher, and Nick Skowronski. My thanks to them all, and absolution for any errors, which are wholly mine. Anyone familiar with the literature will recognize that my contribution lies not in any new information but in my attempt to devise a usable, national context for that existing knowledge.

2. The Pinelands have a rich literature, most of which was summarized shortly after passage of the Pinelands Protection Act (1979). I found three books particularly relevant: Richard T. T. Forman, ed., *Pine Barrens: Ecosystem and Landscape* (New Brunswick, N.J.: Rutgers University Press, 1979); Jonathan Berger and John W. Sinton, *Water, Earth, and Fire: Land Use and Environmental Planning in the New Jersey Pine Barrens* (Baltimore, Md.: Johns Hopkins University Press, 1985); and John McPhee, *The Pine Barrens* (New York: Farrar, Straus, & Giroux, 1967). For terrific distillations of fire history and research, see James A. Cumming, "Prescribed Burning on Recreation Areas in New Jersey: History, Objectives, Influence, and Technique," in *Proceedings, Tall Timbers Fire Ecology Conference 9* (Tallahassee, Fla.: Tall Timbers Research Station, 1969), 251–69, and Kenneth L. Clark, Nicholas Skowronski, and Michael Gallagher, "The Fire Research Program at the Silas Little Experimental Forest, New Lisbon, New Jersey," in press with U.S. Forest Service. The Pinelands Commission and Reserve websites are excellent; see http://www.state.nj.us/pinelands/cmp/summary/. A box score of basic information is available through the New Jersey Pinelands Commission, "Pinelands Facts," July 30, 2012.

3. The best summary of fire history remains Silas Little, "Fire and Plant Succession in the New Jersey Pine Barrens," in Forman, *Pine Barrens*: 297–314.

4. S. Smith quoted in Forman, *Pine Barrens*, 297.

5. See Terry G. Jordan and Matti Kaups, *The American Backwoods Frontier: An Ethnic and Ecological Interpretation* (Baltimore, Md.: Johns Hopkins University Press, 1989).

6. Several excellent summaries exist. See, particularly, Peter O. Wacker, "Human Exploitation of the New Jersey Pine Barrens Before 1900," in Forman, *Pine Barrens*, 3–24, and Berger and Sinton, *Water, Earth, and Fire*, 6–10.

7. Franklin B. Hough, *Report upon Forestry*, 3 vols. (Washington, D.C.: Government Printing Office, 1878), 3:156.

8. Pinchot quote from Gifford Pinchot, "Study of Forest Fires and Wood Protection in Southern New Jersey," *Annual Report of State Geologist for the*

Year 1898, 1898, 11. Chronology from NJFFS website. For an excellent review of the literature prior to 1934, see H. J. Lutz, "Ecological Relations in the Pitch Pine Plains of Southern New Jersey," *Yale University School of Forestry Bulletin,* no. 38 (1934).

9.   Quote from Hough, *Report upon Forestry,* 160.

10.   Granted the significance of the fire, statistics are oddly out of sync. Area burned varies by as much as 20,000 acres, and the number of structures burned by 200. See, for example, James. A. Cumming, "Prescribed Burning on Recreation Areas in New Jersey," 263; Wayne G. Banks and Silas Little, "The Forest Fires of April 1963 in New Jersey Point the Way to Better Protection and Management," *Forest Fire Notes* 25, no. 3 (July 1964): 3–6; David Levinsky, "Remembering Black Saturday: 50 Years Ago, NJ Forest Fires Burned Over 183,000 Acres," *Burlington County Times,* April 22, 2013; Joseph Hughes, "New Jersey, April 1963: Can It Happen Again?," *Fire Management Notes* 48, no. 1 (1987): 3–6.

11.   F. Thomas Ledig and Silas Little, "Pitch Pine (*Pinus rigida* Mill.): Ecology, Physiology, and Genetics," in Forman, *Pine Barrens,* 347–71.

12.   Ralph E. Good, Norma F. Good, and John W. Andresen, "The Pine Barren Plains," in Forman, *Pine Barrens,* 283–95.

13.   Banks and Little, "Forest Fires of April 1963," 6.

14.   Silas Little, "Fire Ecology and Forest Management in the New Jersey Pine Region," *Proceedings, Tall Timbers Fire Ecology Conference 3* (Tallahassee, Fla.: Tall Timbers Research Station, 1964): 35–59; "Effects of Fire on Temperate Forests: Northeastern United States," in T. T. Kozlowski and C. E. Ahlgren, eds., *Fire and Ecosystems* (New York: Academic Press, 1974). On the timeliness of his publications, see Little and E. B. Moore, "Controlled Burning in South Jersey's Oak-Pine Stands," *Journal of Forestry* 43 (1945): 499–506, and S. Little, J. P. Allen, and E. B. Moore, "Controlled Burning as a Dual-Purpose Tool of Forest Management in New Jersey's Pine Region," *Journal of Forestry* 46 (1948): 810–19.

15.   See Clark, Skowronski, and Gallagher, "Fire Research Program," 5–8. In 1985 the Northeastern Research Station (USFS) signed a cooperative agreement with Rutgers University that allowed Rutgers to use the site's buildings for its Pinelands Research Center. There was, apparently, no fire research conducted until the Forest Service reclaimed the facility in 2002.

16.   As an example of this research, see Kenneth L. Clark, et al., "Fuel Consumption and Particulate Emissions During Fires in the New Jersey Pinelands," in *Proceedings of 3rd Fire Behavior and Fuels Conference, October 25–29, 2010* (International Association for Wildland Fire, 2010).

FIRE ON THE MOUNTAIN

1. James Fenimore Cooper, *The Pioneers* (New York: Holt, Rinehart, and Winston, 1967); the fire story unfolds on pages 419–41, from which all quotes are drawn.

2. The footnote can be found at http://www.gutenberg.org/etext/2275, printed on Feedbooks, 418n15.

3. State of New York, *Report of the Forestry Commission*, no. 36 (January 23, 1885): 6. Hereafter, Sargent Report.

4. Sargent, *Report on the Forests of North America*, 9:502.

5. See Louis C. Curth, *The Forest Ranger: A History of the New York State Forest Ranger Force* (New York State Department of Environmental Conservation, 1987), 15.

6. Sargent Report, 9.

7. Sargent Report, 12, 14.

8. Curth, *Forest Ranger*, 16.

9. On wardens and rangers, see Curth, 17.

10. *First Annual Report of the Forest Commission of the State of New York for the Year 1885* (Albany, N.Y.: Argus, 1886), 19–22. *Second Annual Report of the Forest Commission of the State of New York for the Year 1886* (Albany, N.Y.: Argus, 1887), 349–50, 21.

11. *Second Annual Report*, 281.

12. *Sixth Annual Report of the Forest, Fish, and Game Commission* (Albany, N.Y.: James B. Lyon, 1901), 20.

13. *Sixth Annual Report*, 20.

14. *Sixth Annual Report*, 320.

15. *Sixth Annual Report*, 308.

16. H. M. Suter, *Forest Fires in the Adirondack in 1903*, Circular No. 26 (Washington, D.C.: GPO, 1904), 11, 1.

17. For an adept summary see Irland, "Fire History of New York State." For a summary of post-1908 reforms see William G. Howard, *Forest Fires*, Bulletin 10, New York Conservation Commission (Albany, N.Y.: J. B. Lyon, 1914).

18. Roosevelt quoted in Curth, *Forest Ranger*, 24n.

ALBANY PINE BUSH

1. A weird footnote to the pine bush's signature species, the Karner blue butterfly, is that it was first identified by Vladimir Nabokov, the novelist best known for *Lolita* and *Pale Fire*.

2. I was the beneficiary of a marvelous tutorial and field trip organized by Neil Gifford and Tyler Briggs, which included Christopher Hawver, Zak

Handley, and Steve Jackson. My thanks to all of them. They were especially
helpful in tracing out the operational complexities of burning within the
crazy-quilt constraints of the APB.

3.   I've relied on two fundamental references. One is Jeffrey K. Barnes, *Natural History of the Albany Pine Bush: Albany and Schenectady Counties, New York*, New York State Museum Bulletin 502 (New York: New York State Museum, 2003). The other, very rich for its fire theme, is the TNC report: Robert E. Zaremba, David M. Hunt, Amy N. Lester, *Albany Pine Bush Fire Management Plan: Report to the Albany Pine Bush Commission* (New York: New York Field Office, the Nature Conservancy, 1991). I thank Stephanie B. Gebauer for sending me a copy over 20 years ago.

4.   Both quotes from Barnes, *Natural History*, 21–22.

5.   See Barnes, *Natural History*, 22–24, and G. Motzkin, W. A. Patterson III, and D. R. Foster, "A Historical Perspective on Pitch Pine-Scrub Oak Communities in the Connecticut Valley of Massachusetts," *Ecosystems* 2, no. 3 (May–June 1999): 255–73.

6.   Barnes, *Natural History*, 24.

WHERE YOU FIND IT

1.   "Stealthy Underground Fires.; Property on Staten Island's South Shore Endangered," *New York Times*, October 30, 1892, https://www.nytimes.com /1892/10/30/archives/stealthy-underground-fires-property-on-staten-islands -south-shore.html. For data on the 1930 fires, see "Brush Fires Menace Towns on Staten Island and Rage in Wide Suburban Areas," *New York Times*, May 5, 1930, 1. For the "encircling" quote, see "Flames Sweep Forests in East, Houses in Path Are Destroyed; $3,000,000 Fire at Nashua, N.H.," *Star-Gazette* (Elmira, N.Y.), May 5, 1930.

2.   "Our Own Forest Fires," *New York Times*, May 6, 1930, 25.

3.   Peter Charles Hoffer, "New York City's Burning Problem," *New York Times*, April 30, 2006, http://www.nytimes.com/2006/04/30/opinion/nyregion opinions/new-york-citys-burning-problem.html.

4.   Lisa W. Foderaro, "Staten Island Fights Reeds That Feed Its Brush Fires," *New York Times*, April 4, 2012, http://www.nytimes.com/2012/04/05/ny region/a-war-on-reeds-that-feed-staten-island-brush-fires.html.

FOREST AS GARDEN

1.   Alfred Rehder, "Charles Sprague Sargent," *Journal of the Arnold Arboretum* 8, no. 2 (April 1927): 69–86, quote from 69.

2. William Trelease, *Biographical Memoir of Charles Sprague Sargent*, National Academy of Sciences of the United States of America Biographical Memoirs, vol. 12, no. 7 (National Academy of Sciences, 1928), 247–70. For a detailed history of Sprague and the Arboretum, see S. B. Sutton, *Charles Sprague Sargent and the Arnold Arboretum* (Cambridge, Mass.: Harvard University Press, 1970).

3. Trelease, *Biographical Memoir*, 255, 251.

4. Trelease, *Biographical Memoir*, 251.

5. Sargent, *Report on the Forests of North America*, 491. Other quotes from C. S. Sargent, "Forest Fires," in *30th Annual Report of Secretary of Board of Agriculture* (Boston, Mass.: Wright and Potter, 1883), 272–73, 294.

6. Charles S. Sargent, "The Protection of Forests," *North American Review*, 1882, 386, 390, 394.

7. Alfred C. Chapin, *Communication from the Comptroller Submitting Report of the Forestry Commission: State of New York in Assembly, No. 36, January 23, 1885*, 6, 12.

8. For a popular survey of the Commission, see Gerald W. Williams and Char Miller, "At the Creation: The National Forest Commission of 1896–97," *Forest History Today*, Spring/Fall 2005, 32–40. For quote see National Academy of Sciences, *Report of the Committee Appointed by the National Academy of Sciences upon a Forest Policy for the Forested Lands of the United States* (Washington, D.C.: Government Printing Office, 1897), 5.

9. *Report of the Committee Appointed by the National Academy of Sciences upon the Inauguration of a Forest Policy for the Forested Lands of the United States to the Secretary of the Interior, May 1, 1897* (Washington, D.C.: Government Printing Office, 1897), 34.

10. Rehder, "Charles Sprague Sargent," 271–72.

11. Trelease, *Biographical Memoir*, 252.

PITCH PINE AND LEAST TERN

1. I'm indebted to Dave Celino for a wonderful on-site tutorial into Massachusetts fire. Michael Marquardt, District 2 fire warden, contributed insights and documentation. Alex Desrochers and Scott Bauer, firefighters, also assisted.

2. Thomas Morton, *New English Canaan: Or, New Canaan* (1637; repr., New York: Arno Press, 1972), 52–54. A fuller survey is available in Stephen Pyne, *Fire in America* (1982; repr., Seattle: University of Washington Press, 1997), 45–55. I saw no reason to repeat that sweep in this essay, so selected a couple of samples.

3. Sargent, *Report on the Forests of North America*, 500–501.

4. My primary source is the excellent distillation of data in Irland, "Massachusetts Fire History," commissioned by the NFFPC and posted at the North Atlantic Fire Exchange website. Also useful is a PowerPoint presentation from the Massachusetts Bureau of Forest Fire Control, "DCR Fire 2016: Fire Season and Drought," courtesy of Dave Celino.

5. Irland, "Massachusetts Fire History," 14.

6. Irland, "Massachusetts Fire History," 19–20; "Fire Season in Plymouth and Barnstable Counties" and "Narrative Report: Carver-Wareham-Plymouth Fire of May 23 and 25, 1964," unpublished reports, n.d., courtesy of Massachusetts DCR.

7. Figures from "DCR Fire 2016: Fire Season and Drought," slides 35–36.

THE WUI WITHIN

1. For their marvelous tutorial I would like to thank the entire staff of the NFPA's Wildland Fire Operations Division, especially Michele Steinberg and Molly Mowery, along with Sue Marsh and Jessica Broady of the library and archives, and James Shannon, president and CEO.

THE VIEW FROM BILL PATTERSON'S STUDY

1. Unless otherwise indicated, information comes from an interview with Bill Patterson on April 30, 2017.

2. H. Ali Crolius, "Prometheus's Gift," *UMass Magazine*, Fall 1998, 3, http://www.umass.edu/umassmag/archives/1998/fall_98/fall98_f_fire.html.

3. Email from William A. Patterson III to Stephen Pyne, April 1, 2017.

4. Email from William A. Patterson III to Stephen Pyne, April 26, 2017.

5. See Motzkin, Patterson, and Foster, "Historical Perspective," 255–273.

6. Crolius, "Prometheus's Gift," 3.

MAINE'S EPICYCLES OF FIRE

1. See Charles B. Fobes, "Historic Forest Fires in Maine," *Economic Geography* 24, no. 4 (October 1948): 269–73. Most comprehensive summary, drafting heavily from biennial reports from the Forest Commission, is available in Philip T. Coolidge, *A History of the Maine Woods* (Bangor, Maine: Furbush-Roberts, 1963).

2. On the 1825 fire, see *Second Annual Report of the Forest Commissioner of the State of Maine, 1894* (Augusta, Maine: Burleigh and Flynt, 1894): 37–40. On the Katahdin fire, Coolidge, *History*, 128.

3.	Coolidge, *History*, 127–35. His major source is the biennial reports of the Forest Commissioner of Maine, particularly the report for 1894.

4.	Coolidge, *History*, 135–40.

5.	Sargent, *Report on the Forests of North America* (Exclusive of Mexico), 494.

6.	Quote from Coolidge, *History*, 166. Commission quote from *Fifth Report of the Forest Commissioner of the State of Maine, 1904* (Augusta, Maine: Kennebec Journal Print, 1904), 7. A special report by Springer gives a warden-by-warden report on the year's fires.

7.	Dana quoted in Coolidge, *History*, 171. For a detailed study of fires in an interesting decade, after the Great Fires, but before the CCC and the rapid decline in starts and area burned, see Samuel T. Dana, "Forest Fires in Maine, 1916–1925," *Bulletin No. 6*, Maine Forest Service (Augusta, 1926).

8.	Austin H. Wilkins, *Ten Million Acres: The Remarkable Story of Forest Protection in the Maine Forestry District (1909–1972)* (Woolwich, Maine: TBW Books, 1978), 192–93.

9.	Joyce Butler, *Wildfire Loose: The Week Maine Burned* (Kennebunkport, Maine: Durrell, n.d.); A. G. Hall, "Four Flaming Days," *American Forests*, December 1947; Austin Wilkins, "The Story of the Maine Forest Fire Disaster," *Journal of Forestry* 46 (1948): 568–73.

10.	The best description of what happened are the statistical records gathered and analyzed by Lloyd C. Irland: "Maine Forest Fire History and Analysis 1903–2010, Working Paper, Revised Draft June 2012" and "Working Paper, Maine Regional and Individual Fire Data 1903–2010," both made available by the author. Lloyd, I'm grateful.

11.	On Baxter State Park, see John W. Hakola, *Legacy of a Lifetime: The Story of Baxter State Park* (Woolwich, Maine: TBW Books, 1981).

12.	I'm indebted to Joe Mints of the MFS for insight into the possible role of the 1903 burn for the park's history.

13.	Fitzgerald v. Baxter State Park Auth., 385 A.2d 189 (1978), https://law.justia .com/cases/maine/supreme-court/1978/385-a-2d-189-0.html.

14.	Data from Board of Review, "Baxter Park Fire, Old Town Forestry Headquarters, August 24, 1977." Copy provided by Maine Forest Service.

15.	Several accounts: From those advocating wilderness values, see Aimee Gaudin, "As Baxter Park Burns, So Burns Maine," *Audubon* 79 (January 1977): 146–53. For those advocating active intervention, see Vladek Kolman, "The 1977 Baxter State Park Fire," *Forester*, 1978, 25–28, https:// library.umaine.edu/forester/content/Forester_1978_027–049.pdf. Note that Kolman was the owner of the firm hired by the park to do restoration work before the fire broke out. For the context within the park's history of management, see Hakola, *Legacy*, 304–05.

16. On the 1952 and 1953 lightning fire seasons, see Coolidge, *History*, 164, who relies on the biennial reports, and Charles B. Fobes, "Lightning Fires in the Forests of Northern Maine, 1926–1944," *Journal of Forestry* 42 (April 1944). For a counterview of lightning, more in alignment with boreal forests generally, see Timothy J. Fahey and William A. Reiners, "Fire in the Forests of Maine and New Hampshire," *Bulletin of the Torrey Botanical Club* 108, no. 3 (July–September 1981): 362–73.

17. The site underwent several studies at the time TNC acquired it: Polly Harris, "Waterboro Barrens: A Report Prepared by Polly Harris for the Nature Conservancy, November 1991"; William A. Patterson III, "The Waterboro Barrens: Fire and Vegetation History as a Basis for the Ecological Management of Maine's Unique Scrub Oak-Pitch Pine Barrens Ecosystem: A Report Submitted to Maine Chapter, the Nature Conservancy, February 1, 1994"; Carolyn Ann Copenheaver, "Determinants of Vegetation Distribution in a Southern Maine Pitch Pine-Scrub Oak Ecosystem" (master's thesis, University of Maine, 1996).

I'm grateful to Nancy Sferra for sending me copies and for arranging a marvelous field tutorial at Waterboro Barrens and Wells Barrens Preserves. I also thank the other participants for contributing their time and knowledge: Jon Bailey, Dan Grenier, Andy Cutko, Tyler Cross, Fran Riesner, William Brewer, Birch Milotky, and Kassandra Strango.

I'm also indebted to Jeff Currier, Joe Mints, John Crowley, and William Hamilton of the Maine Forest Service for an informative conversation about the agency's past and future, for some useful documents, and for getting me around Baxter State Park. Jeff, in particular, sharpened my understanding of what Maine's demographic changes mean for fire management.

COLLECTIVE SECURITY

1. Austin Wilkins quoted in Butler, *Wildfire Loose*, 221. Butler's book is most comprehensive. For a thumbnail summary, see the New England Historical Society's webpage on the fire at http://www.newenglandhistoricalsociety .com/the-year-a-state-burned-maine-fires-of-1947/.

2. USFS quote from Butler, *Wildfire Loose*, 221.

3. Lloyd C. Irland, "Northeast Forest Fire Protection Compact: Stress-Testing Study" (presentation, NE Compact Conference, Plymouth, Mass., August 3, 2011), and Irland, "Stress Testing Project Update" (Compact Winter Meeting, South Portland, Maine, January 24, 2012). The project involved

a general data sweep of the members—all in all, an extraordinary resource for researchers.

WESTWARD, THE COURSE OF EMPIRE

1. Ernest Hemingway, "The Big Two-Hearted River," in *The Short Stories of Ernest Hemingway* (New York: Scribner, 1953), 209.
2. Hemingway, "Big Two-Hearted River," 212, 211.
3. Hemingway, "Big Two-Hearted River," 232.
4. The best overview remains Susan Flader, ed., *The Great Lakes Forest: An Environmental and Social History* (Minneapolis: University of Minnesota Press, 1983).
5. Quotes from Pyne, *Fire in America* (1982), 201, 210–11.
6. Among the standard accounts see Rev. Peter Pernin, "The Great Peshtigo Fire," *Wisconsin Magazine of History* 54 (1971): 246–72; Denise Gass and William Lutz, *Firestorm at Peshtigo: A Town, Its People, and the Deadliest Fire in American History* (New York: Henry Holt, 2002); Lawrence H. Larsen, *Wall of Flames: The Minnesota Forest Fire of 1894* (Fargo: North Dakota Institute for Regional Studies, North Dakota State University, 1984); Grace Stageberg Swenson, *From the Ashes: The Story of the Hinckley Fire of 1894* (Stillwater, Minn.: Croixside Press, 1979); and Francis M. Carroll and Franklin R. Raiter, *The Fires of Autumn: The Cloquet-Moose Lake Fire of 1918* (St. Paul: Minnesota Historical Society Press, 1990). Among the new contenders is Daniel James Brown, *Under a Flaming Sky: The Great Hinckley Firestorm of 1894* (Lanham, Md.: Rowman and Littlefield, 2009).
7. Jim Harrison, *True North* (New York: Grove/Atlantic: 2005), 23.
8. A classic account of how the logging was done is available Agnes M. Larson, *The White Pine Industry in Minnesota: A History* (Minneapolis: University of Minnesota Press, 1949).
9. A good survey of state efforts is Raleigh Barlowe, "Changing Land Use and Policies: The Lake States," in *The Great Lakes Forest*, ed. Susan Flader (Minneapolis: University of Minnesota Press and Forest History Society, 1983), 156–76.
10. The sad story is nicely told in Curt Meine, *Aldo Leopold: His Life and Work* (Madison: University of Wisconsin Press, 1988), 518–20.
11. Rodney W. Sando and Donald A. Haines, *Fire Behavior of the Little Sioux Fire*, Research Paper NC-76, U.S. Forest Service, 1972, and Albert J. Simard et al., *The Mack Lake Fire*, General Technical Report NC-83, U.S. Forest Service, 1983.

12. See Seney National Wildlife Refuge, https://www.fws.gov/refuge/seney /what_we_do/history.html.

13. I rely on my account in *Between Two Fires: A Fire History of Contemporary America* (Tucson: University of Arizona Press, 2015), 152–53.

14. Several studies exist. A useful summary is available in Dawn S. Marsh, "The Evolution of Land and Fire Management at Seney National Wildlife Refuge: From Game to Ecosystem Management," report for Fish and Wildlife Service, December 2013.

EPILOGUE

1. See, for example, L. M. Nagel et al., "Adaptive Silviculture for Climate Change: A National Experiment in Manager-Scientist Partnerships to Apply an Adaptation Framework," *Journal of Forestry* 115 (2017): 167–78.

INDEX

ABOUT THE AUTHOR

Stephen J. Pyne is an emeritus professor at Arizona State University and a former North Rim Longshot. Among his recent books are *Between Two Fires: A Fire History of Contemporary America* and To the Last Smoke, a series of regional fire surveys. He lives in Queen Creek, Arizona.